Exergy - Theoretical Background and Case Studies

Edited by Joan Josep Roa Rovira

Published in London, United Kingdom

Exergy - Theoretical Background and Case Studies
http://dx.doi.org/10.5772/intechopen.1001646
Edited by Joan Josep Roa Rovira

Contributors
Alfredo Constaín Aragón, Ali Raza, Etienne Tchoffo Houdji, Farzaneh Hatami Kamin, Laura Felício, Liaquat Ali Khan, Marva Hadia, Miguel St. Aubyn, Naradasu Ravi Kumar, Pasupuleti Ravindra Kumar, Rodrigo Bulnes A., Shahriar Salimi, Steven Audrey Heugang Ndjanda, Teles Huo, Tânia Sousa, Zunaira-Tu-Zehra

First published in London, United Kingdom, 2024 by IntechOpen
IntechOpen is the global imprint of INTECHOPEN LIMITED, registered in England and Wales, registration number: 11086078, 167-169 Great Portland Street, London, W1W 5PF, United Kingdom

British Library Cataloguing-in-Publication Data
A catalogue record for this book is available from the British Library

Additional hard and PDF copies can be obtained from orders@intechopen.com

Exergy - Theoretical Background and Case Studies
Edited by Joan Josep Roa Rovira
p. cm.
Print ISBN 978-0-85466-683-6
Online ISBN 978-0-85466-682-9
eBook (PDF) ISBN 978-0-85466-684-3

For EU product safety concerns:
IN TECH d.o.o., Prolaz Marije Krucifikse Kozulić 3, 51000 Rijeka, Croatia,
info@intechopen.com or visit our website at intechopen.com.

Meet the editor

Dr. Joan Josep Roa is the Test Laboratory Manager at STEROS GPA INNOVATIVE S.L.U. He earned a degree in chemical engineering from the University Rovira I Virgili, Spain, in 2005, followed by a degree in materials engineering from the Universidad Politécnica de Cataluña (UPC), Spain, in 2007. He obtained a Ph.D. from the University of Barcelona, Spain, in 2010, after which he pursued a two-year postdoctoral fellowship at the "Centre National de la Recherche Scientifique," France. In 2012, Dr. Roa returned to UPC with a Juan de la Cierva contract, funded by the Ministry of Economy and Competitiveness. In 2016, he became a tenure-track associate professor in the Department of Materials Science and Metallurgy through the Serra Húnter Programme. Since 2020, he has served as an associate professor at UPC. Dr. Roa's research involves the mechanical characterization of materials at different length scales, characterization of plastic deformation mechanisms, 3D printing technology, and development and characterization of conductive polymeric materials. He has co-authored more than 160 peer-reviewed publications, 6 reviews, 9 book chapters, 1 book, and 240 proceedings and reports primarily focused on advanced mechanical characterization techniques. He has presented more than 225 communications, including 20 invited talks. He is also a co-creator of patent applications and has contributed to various industrial implementations. Additionally, he is responsible for managing both international and national R&D projects.

Contents

Preface

The field of energy analysis has long been dominated by the concept of energy conservation, a principle rooted in the First Law of Thermodynamics. However, while energy cannot be created or destroyed, it can undergo transformations that render it less useful for performing work, a limitation that traditional energy analysis often overlooks. This realization has led to the emergence and growing importance of exergy analysis, which provides a more comprehensive understanding of energy systems by accounting for the quality of energy, not just its quantity. Exergy, a concept derived from the Second Law of Thermodynamics, measures the maximum useful work that can be obtained from a system as it moves towards thermodynamic equilibrium with its environment. Unlike energy, exergy is destroyed in irreversible processes, making it a more insightful tool for evaluating the efficiency of energy conversion systems. By focusing on the degradation of energy quality, exergy analysis helps identify where and how losses occur, guiding improvements in both process design and resource management.

The theoretical background of exergy is rooted in classical thermodynamics, but its application extends far beyond traditional boundaries. It provides a rigorous framework for analyzing the efficiency of mechanical and thermal systems in terms of exergy, allowing researchers and engineers to better understand the true work potential within a system and identify opportunities for minimizing waste, thereby contributing to more sustainable practices. This book underscores the importance of exergy in contemporary scientific and engineering applications, exploring its theoretical foundations and practical implications. As the world grapples with the twin challenges of energy sustainability and environmental impact, exergy analysis emerges as an essential tool for designing systems that are not only more efficient but also more aligned with the principles of sustainability. By providing a clearer picture of how energy is utilized and degraded, exergy analysis opens new pathways for innovation in energy technology, ultimately contributing to the creation of more resilient and sustainable energy systems.

Dr. Joan Josep Roa Rovira
Steros GPA Innovative S.L.,
Barcelona, Spain

Section 1

Introduction to Exergy

Chapter 1

Challenges in Assessing Useful Exergy and Its Significance for Economic Development Studies: An Examination of Mozambique's Economics Sectors, 1971–2014

Teles Huo, Miguel St. Aubyn, Laura Felício and Tânia Sousa

Abstract

This chapter presents and analyses the outcomes derived from estimating useful exergy data associated with the final electricity usage in Mozambique's productive sectors—encompassing industry, commerce, and services—covering Mozambique, from 1971 to 2014. At the time of carrying out the research, the availability of International Energy Agency (IEA) data relating to Mozambique was only from 1971 to 2014. The societal exergy analysis methodology used to estimate useful exergy involved first converting the final electricity use into its exergetic equivalent. Then, this exergy was categorized according to its specific end uses. Finally, the useful exergy was calculated by applying the appropriate efficiency rates for each usage category. The electricity consumption data, measured in GWh, was sourced from the IEA database. The findings indicate a significant increase in final electricity consumption starting in 2000. While there was a noticeable rise in useful exergy in the trade and services sector, the industry sector's contribution remained dominant. The most significant contributions to total useful exergy were mechanical work in the industry sector, followed by high-temperature heat production. As useful exergy represents the effective portion of exergy reflecting energy usage outcomes, it emerges as a crucial concept for analyzing the link between energy consumption and economic growth.

Keywords: economic growth, energy, exergy, final electricity consumption, Mozambique, useful exergy

1. Introduction

While mainstream economics traditionally emphasizes capital and labor as the primary factors of production, this chapter contends that energy and exergy are equally vital for the production process and economic growth analyses. This perspective aligns with the first two laws of thermodynamics asserting that in all real processes, energy is conserved (first Law) but degraded (second Law). The degradation of energy can be measured as a loss of its ability to perform mechanical work, commonly referred to as exergy [1].

Considerable evidence supports the contention that energy is crucial for economic growth, and in the production process, there is no substitute for energy. Although different energy sources can be substituted, energy as a whole cannot. Hence, energy is as integral to the production process as capital and labor. Without energy, generating a product is impossible, making energy more than a mere component of production costs; it is an intrinsic part of the production process. Analyzing the relationship between energy consumption and economic dynamics is more effective when using useful exergy, that is. the exergy of the heat produced by the electric radiator or mechanical work produced by the electric motor, compared to final electricity consumption (FEC) [2, 3].

Despite the significance of energy, there is a notable absence of exergy studies, particularly in African countries. No useful exergy research related to Mozambique has been found [2, 3]. Concerning Africa, the research conducted by Ref. [4] focused on determining the exergetic efficiencies of electricity and fuel consumption in South Africa, examining various sectors and overall usage. Their study estimated that the exergetic efficiency of the mining industry, the most dominant sector in South Africa, is approximately 83%. Ultimately, the researchers argue that the performance of exergy utilization in the South African industry can be enhanced by adopting different forms of conservation. Turning to Ref. [5], their analysis encompasses Ghana and the United Kingdom. They examine the relationship between the imperative to reduce energy consumption and Gross Domestic Product (GDP) growth (absolute decoupling), considering climate change objectives by 2030. The researchers conclude that achieving "absolute decoupling" between primary energy consumption and economic activity will be very challenging for both Ghana, as a growing economy requiring more energy to sustain its growth, and the United Kingdom, where energy consumption is expected to remain constant in the future to ensure GDP growth. Additionally, they assert that energy efficiencies alone are not effective in reducing primary energy consumption or carbon dioxide (CO2) emissions.

Regarding global studies on exergy, Ref. [6] analyses the exergetic efficiencies inherent to the different forms of exergy use while [7] reconstructs the different forms of electricity use in the United States of America (USA) economy, from 1900 to 1998, using different data sources. The electricity uses were classified into: lighting, heating (high and low temperatures), mechanical work, electrolysis, and electronic devices. Results showed that the USA largest consumption was for mechanical work and that the final to useful aggregated exergy efficiency of electricity remained unchanged during the twentieth century. This is consistent with results obtained by Ref. [8] that show that despite technological evolution, the World final-to-useful end-use efficiency was surprisingly constant (~48%), due to "efficiency dilution," wherein individual end-use efficiency gains are offset by increasing uptake of less efficient end uses.

The study [9] discusses the usefulness of the concept of exergy for all systems subject to mass and energy flows and argues that economic and ecological systems are highly coupled, given the possibility of irreversible depletion of the available planetary stocks of exergy.

This chapter aims to present and analyses the results and constraints of constructing useful exergy data associated with the FEC of Mozambique's economic productive sector, encompassing industry, commerce, and services from 1971 to 2014. This period is imposed by the availability of IEA data relating to Mozambique, which at the time of carrying out the research, were only from 1971 to 2014.

These data hold significance for studying the role of useful exergy in Mozambique's economic growth, thereby contributing to the ongoing debate on this subject.

Economic growth, quantified by the evolution of GDP, is not solely dependent on input availability but also on effective work derived from FEC. Consequently, from an engineering standpoint, production—whether industrial or otherwise—requires energy-driven work [1].

In the production process, useful exergy is essential because work transforms inputs into final products. Useful exergy finds applications in various sectors, such as industry, where it powers machines through mechanical drive, involving the conversion of electricity into mechanical energy [10]. Additionally, it plays a crucial role in heating processes, such as those employed in smelting or firing processes within heater systems.

Addressing the relevance of energy in Mozambique's economy, the Industrial Policy and Strategy for 2016–2025 (IPS) identifies poor electricity access in many regions, particularly outside urban areas, and high electricity costs as key structural constraints for industrial development in the country.

Following this introduction, the second section delves into the concept of useful exergy, the third section presents data and methods, the fourth section outlines the results, and the final section encompasses the discussion and conclusions.

2. From exergy to useful exergy

Exergy is the useful part of the energy of a system in a specific environment, that is. the maximum amount of work, whether mechanical, heat generation, lighting, or other, that a system can perform until achieving equilibrium with its environment [10].

For a closed system in steady state, the exergy balance corresponds to the Eq. (1) [4]:

$$\sum_r E^{Q^r} - E^w - I = 0 \tag{1}$$

Where:

E^{Q^r} and E^w indicate the values of exergy associated with heat transfer (Q^r) and the performance of the work (W) provided by the system.

Also, according to Ref. [4], when the heat transfer (Q^r) occurs in a space with uniform temperature, T_r, and T_o, and the environment temperature[1], then the associated thermal exergy is given by:

$$E^{Q^r} = \left(1 - \frac{T_0}{T_r}\right).Q^r \tag{2}$$

On the other hand, the exergy of mechanical work is expressed as being [4]:

$$E^w = W \tag{3}$$

The exergy flows, E^{Q^r} and E^w, measure the quality of the energy flow in its state of use, akin to an energy service, resulting in what is commonly referred to as "useful work" [1].

I denotes the exergy consumed or lost due to irreversibility, conferred by the second law of thermodynamics.

In a real process, the exergy destruction is the quality of energy that is destroyed, which is proportional to the entropy production [12]. Therefore, the consumption of exergy is greater than zero for any irreversible process and equal to zero for reversible processes.

The idea of exergy as the maximum amount of work that can be produced from a given energy flow is conceptually shared among exergy authors [3, 13], and so it will be understood in this chapter, as the part of the energy used to generate work during production processes, while the useful exergy is obtained as a result of applying the conversion efficiencies related to each end use.

3. Data and methods

The methodology applied for the construction of useful exergy series is the societal exergy analysis [14]. The FEC data are in GWh. The data are from IEA database, covering the period of 1971 to 2014, by sectors of activity, excluding, in this chapter, the residential sector. There is no data related to the Mozambican agricultural sector.

The total FEC trend, in **Figure 1**, indicates that there are two distinct periods: first, from 1971 to 1999 and, second, from 2000 to 2014. The second period is strongly influenced by the beginning of MOZAL's activities in Mozambique, an industry of aluminum smelter with final electricity consumption corresponding to 45% of the country's consumption.

[1] For Mozambique, the reference for the average temperature was obtained from [11], indicating that the annual average temperature varies from 22 to 24°C. The Maputo region records high average temperatures (24°C) when compared to other regions in the country. Furthermore, Maputo hosts the country's largest industrial park, approximately 55% of all industries in the country [11]. For that reason, the annual average temperature of Maputo was taken as a reference for the environment temperature.

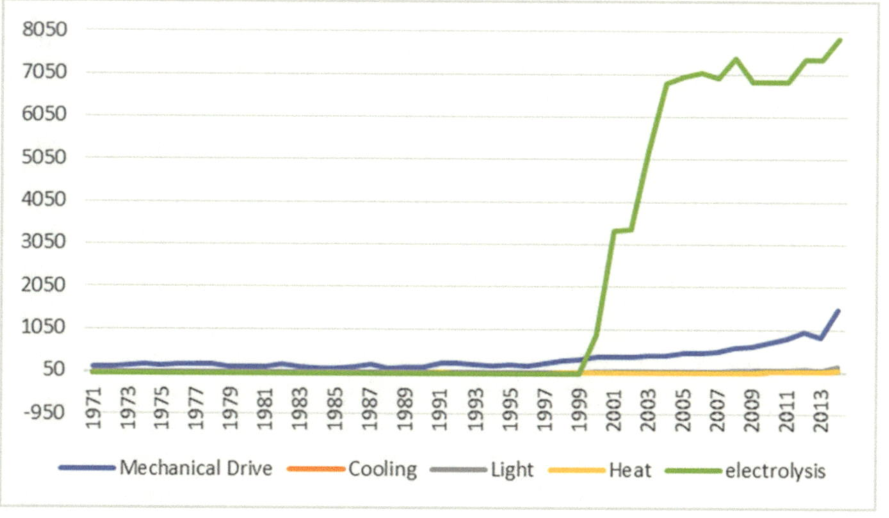

Figure 1.
Exergy by end uses: Industrial sector.

The methods followed three steps: (1) the conversion of FEC into exergy, (2) the allocation of final exergy consumption, per sector, to the different end uses, and (3) the application of the final exergy efficiencies, according to end uses, to obtain the useful exergy data per sectors for Mozambique. The total useful exergy data was obtained by summing the disaggregated results.

Exergy conversion factors were applied to obtain exergy data. In the case of electricity, the conversion factor to exergy is equal to one.

The end uses for industrial, commercial and services sectors followed the end uses obtained from the literature [7, 15, 16], which indicate types of end uses, such as mechanical drive, heat, lighting, cooling, and other uses.

In the case of industry, five end uses have been identified: the use of electricity for mechanical work, heat production, lighting, cooling, and electrolysis processes. For the commerce and services sector, four use-ends were found for lighting, cooling, heat, and electronics uses.

Useful work, which is useful exergy, is calculated as [12]:

$$U_{tjk} = \varepsilon_{tk} \phi E_{tjk} \tag{4}$$

Where:

j—represents economic sectors,

k—indicates the category of usage types,

t—is the time,

ε_{tk} —are the exergy efficiencies by usage types;

ϕ —denotes the exergetic factor for each type of energy source (the exergetic factor is equal to 1 for electricity), and.

E_{tjk} —is the exergy obtained by type of use for each production sector.

Considering the specified end uses for each sector, it is essential to apply the corresponding exergy efficiencies to derive the useful exergy for each sector. Examining mechanical drive, historical data indicates that in 1970, industrial engines demonstrated an average efficiency of approximately 70%. Subsequent advancements led to an increased average efficiency, reaching around 85% in 1980 and further climbing to 90% in the 2000s [7].

A comprehensive investigation into lighting within the US industrial sector in 1946 disclosed that lighting efficiencies accounted for about 7.7% of industrial electricity consumption at that time [7]. The efficiency of lighting is contingent upon the type of lamp used. In 1970, the efficiency for incandescent lamps was 5%, while that for fluorescent lamps was 20% [7]. Determining the applicable efficiency should consider the predominant lamp type in use within production units. However, in many instances, data related to the predominant lamp types in different economic sectors is not available. Over time, lighting efficiencies exhibited notable improvements from 7% in 1940 to 8.5% in 1960, 11% in 1980, and further advancing to 13% in 2000 [7].

In the context of cooling associated with air conditioning, the efficiency is estimated to be around 0.3 [8]. Refrigeration, on the other hand, experienced a notable efficiency increase from 75% in 1960 to 85% in 2000, attributed to advancements in equipment enhancing overall efficiency [7]. The efficiency of heat at high temperatures also underwent changes over time, progressing from 87% in the 60s and 70s to 89% in 1980, culminating in 90% by 2000 [7].

Heat exergy efficiencies are calculated using Carnot efficiency. The exergetic efficiency of heat production depends on the environment and the system temperature [15]. For heat production, three subcategories were established giving the system temperature: High-temperature heat (HTH), for temperatures above 500°C (industrial uses); medium temperature heat (MTH), for temperatures between 500°C and 120°C; and Low-temperature heat (LTH), for uses below 120°C [15]. LTH uses can be further disaggregated into three other sub-subcategories: LTH 1, for uses between 120°C and 90°C; LTH2, for uses between 90°C and 50°C; and LTH3, for uses below 50°C. In this study, LTH2 was assumed for commerce and service sectors [15]. For LTH, electric heat has an exergetic efficiency of around 10% [10].

In this study, the efficiencies adopted were adapted from Ref. [15] for mechanical work, heat production, lighting, cooling, and others, as discussed in point 4 of this paper. The efficiencies applied are part of Appendices A and B.

4. Results

4.1 Exergy allocations by end uses

Results of exergy by end uses are shown in **Figures 1** and **2**.

Figure 1 shows the exergy allocated by end uses in the production processes of industrial sector, including mechanical drive, applied for running machines, cooling, lightning, heat and electrolysis processes applied, especially in aluminum industries..

Up until 1999, the predominant utilization of exergy in the industrial sector was in mechanical drive. However, starting from 2000 onwards, the most significant application shifted to electrolysis, corresponding to the introduction of the aluminum industry, particularly with the establishment of MOZAL. In contrast, for the commerce and service sector, among the end uses found in this sector, the primary use throughout the entire period remained cooling, as depicted in **Figure 2**.

Figures 1 and **2** illustrate the exergy uses in both sectors, giving the pattern of uses according to the energy needs of each sector.

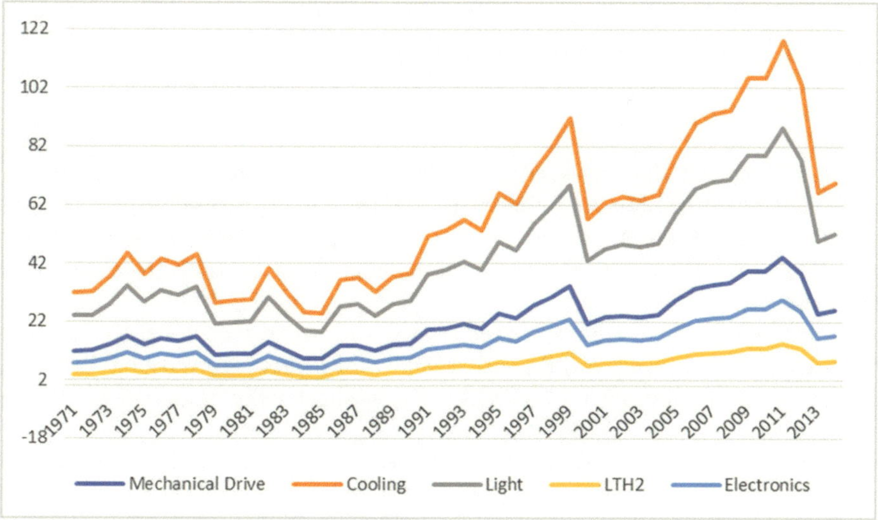

Figure 2.
Exergy by end uses: Commercial and service sector.

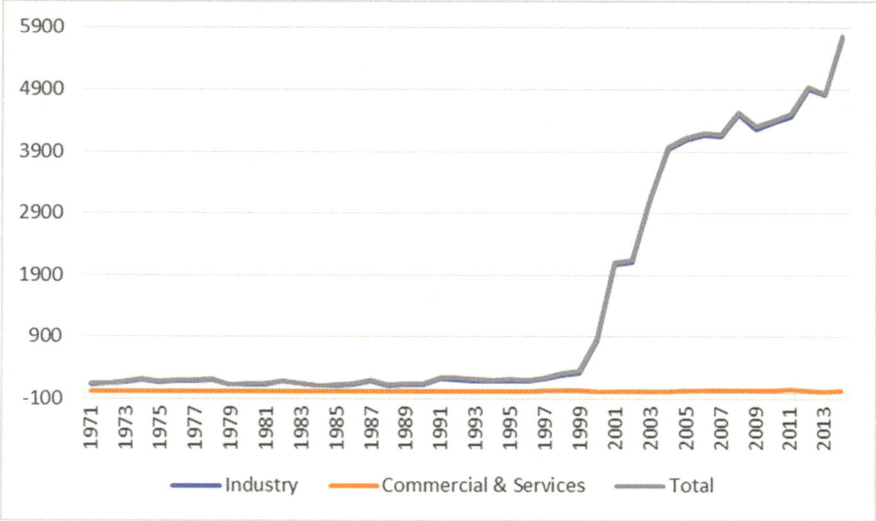

Figure 3.
Useful exergy by sector and total.

The next stage was to obtain the useful exergy data by applying exergy efficiencies related to each type of end uses. Details about the efficiencies applied are in Appendices A and B at the final of the chapter..

4.2 Useful exergy data

Figure 3 shows the useful exergy data obtained for industrial and for commercial and services, including the total useful exergy of both sectors.

The findings indicate that the total useful exergy is primarily influenced by the useful exergy generated within the industrial sector.

4.2.1 Total useful exergy

Figure 4 illustrates the total useful exergy, revealing a dynamic pattern akin to that of FEC. This resemblance can be attributed to the diminished impact of heat end uses in Mozambique. As previously highlighted, the conversion of electricity to exergy follows a nearly one-to-one ratio for various end uses, with the exception of processes involving heat.

From 2000 onwards, the energy needs of the industrial sector arose steadily as a result of the sector recovery and the effectiveness of direct international investment, like that of the implementation of MOZAL, accrued after the cease-fire in 1992, and after the implementations of the first democratic government.

Figure 5 provides a detailed depiction of the useful exergy within the commercial and service sector. The findings reveal a seasonal dynamic in this sector from 1971 to 1991, followed by a steady growth trend from 1992 until 2011. This upward trajectory corresponds to the postwar period, specifically after the conclusion of the conflict in 1992. Conversely, the decline in 2000 is associated with the occurrence of a significant flood that affected the economy.

4.2.2 Efficiency of useful exergy

Figure 6 illustrates the overall efficiency of useful exergy. The efficiency of useful exergy represents the ratio of useful exergy to FEC (both total and by sectors). This ratio serves as an indicator of the efficient utilization of the FEC.

The results indicate that from 2000 onwards, despite the increase in total useful exergy, the overall efficiency of useful exergy was lower than in the previous period. This is attributed to the substantial contribution of electrolytic uses during this period, which exhibit lower efficiencies compared to mechanical drive, the dominant

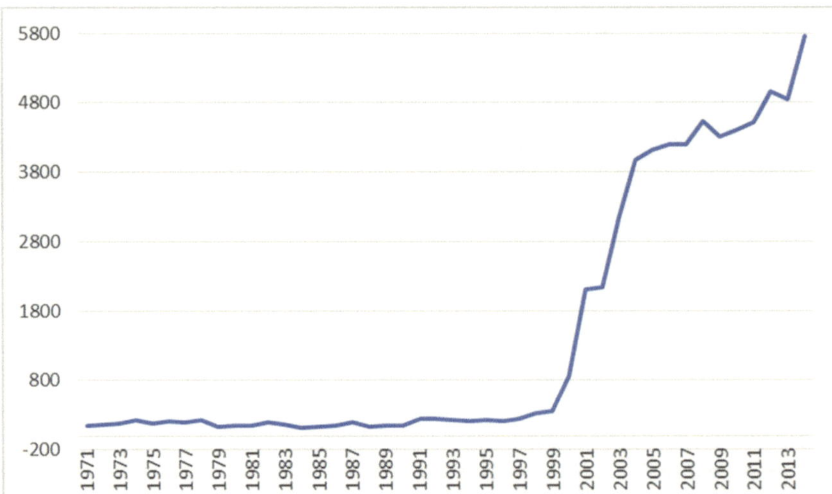

Figure 4.
Total useful exergy.

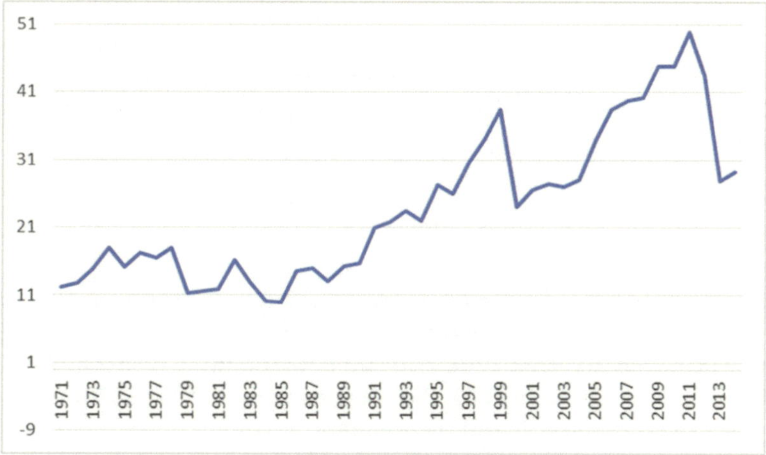

Figure 5.
Useful exergy of commercial and services sector.

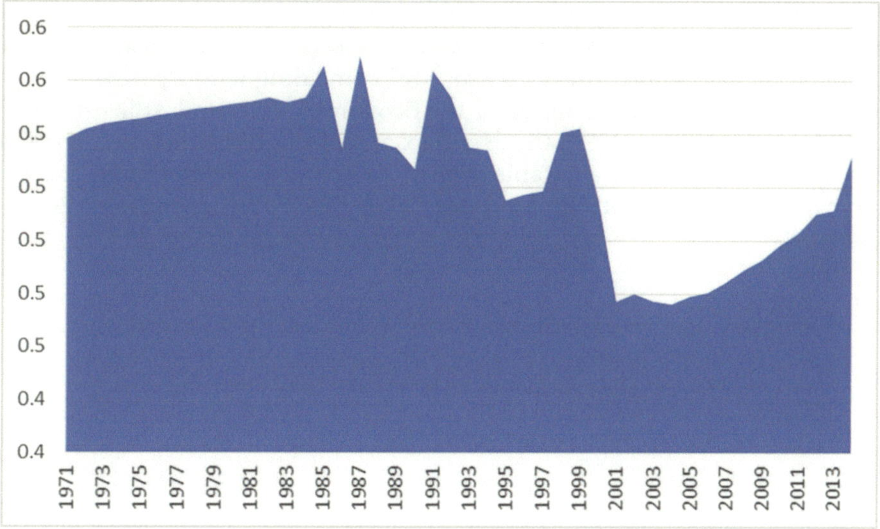

Figure 6.
Total efficiency of useful exergy.

factor in the preceding period. Below are the total efficiencies of useful exergy categorized by sectors: Industrial, commercial, and services sectors [2]:

[2] In Mozambique, the industrial sector is predominantly composed of micro- and small-sized enterprises, representing 90% of the sector. Micro-industries make up 63% of the sector, small ones account for 31%, medium-sized enterprises constitute 3%, and large industrial companies, with the potential for higher electricity consumption, represent only 3%. However, despite being few in number, large industrial companies are the major employers, employing 71% of the workforce [17]. Regarding the industrial sector's contribution, the metallurgical sector contributes 35% to industrial production, the food sector contributes 25%, beverages contribute 13%, nonmetallic minerals contribute 10%, tobacco contributes 8%, and the remaining sectors contribute 9% to the total industrial production [17]. The other sector is commercial and services, with less electricity consumption patterns, compared to the industrial sector.

Figure 7, depicting the total efficiency of useful exergy by sectors, reveals a notably high efficiency for the industrial sector. This is attributed to the substantial utilization of both FEC and useful exergy within that sector, as compared to the commercial and service sector. The disparities in energy demand between the two sectors also contribute to this variation.

5. Conclusions

The total amount of useful exergy remained nearly constant from 1971 until 1990, experiencing moderate growth between 1991 and 1999. Subsequently, an explosive growth phase occurred between 2000 and 2004, maintaining an upward trajectory until 2013. A brief stagnation in 2001 and 2002 is attributed to the 2000 floods, impacting the productive sector and the economy at large.

The total useful exergy is strongly influenced by the industrial sector. Although the useful exergy of the commerce and services sector has shown modest growth, the total useful exergy had a significant growth induced by the growth of the useful exergy of the industrial sector.

However, a closer examination of the commerce and services sector reveals a change in dynamic trends over time. From 1971 to 1978, there was an increasing trend, followed by a decline from 1979 to 1985, coinciding with a period of economic crisis and the introduction of the Economic Structural Adjustment Plan in 1987. From 1986 to 1999, the useful exergy of the commerce and services sector grew rapidly. Despite a notable drop in 2000 due to severe floods, there was a recovery in 2001. However, a substantial decline occurred between 2012 and 2013.

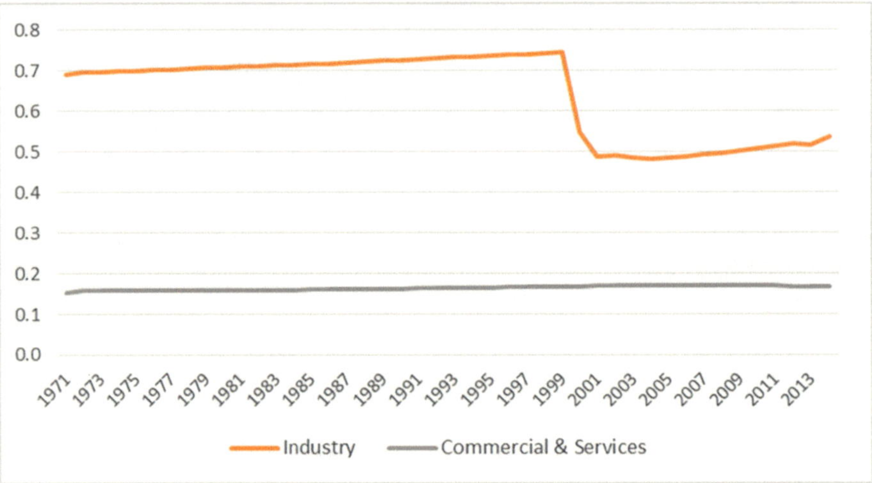

Figure 7.
Total efficiency of useful exergy by sectors.

Regarding end uses, the industry sector's mechanical work made the greatest contribution to total useful exergy. Post-2000, the total efficiency of useful exergy was lower than in the preceding period.

Several limitations affect this analysis, including a lack of detailed data on various economic sectors, especially the absence of data on FEC uses and the agriculture sector for Mozambique. Additionally, detailed efficiencies associated with different end uses suitable for the technological stage of underdeveloped countries are unavailable. Despite these constraints, this study serves as a foundational point for future research, incorporating diverse energy sources for determining the country's total useful exergy.

In conclusion, useful exergy, as an indicator of the outcome of energy use, holds significance for analyzing the nexus between energy consumption and economic growth. Policymakers must prioritize providing records of energy end uses and exergy efficiencies sensitive to the technological stage of underdeveloped countries, such as Mozambique. This is crucial for conducting more precise and accurate analyses of the dynamic relationship between energy and economic factors.

Acknowledgements

The authors wish to thank Tiago Domingos, from IST - Lisbon University, for his invaluable suggestions.

This work was supported by the FCT (Fundação para a Ciência e a Tecnologia) [grant number UIDB/05069/2020].

Conflict of interest

The authors declare no conflict of interest.

Acronyms

FEC	Final electricity consumption
GDP	Gross domestic product
CO_2	Carbon dioxide
IEA	International energy agency
ISP	Industrial policy and strategy
HTH	High-temperature heat
LTH	Low-temperature heat
USA	United States of America

Appendix A

See **Table A1**.

Year	Electrolysis (%)	Mechanical drive (%)	Cooling (%)	Light (%)	Heat (%)	HTH (aluminum) (%)
1970	32.41	83.45	11.31	0.90	32.37	62.51
1971	32.66	83.67	11.29	1.14	32.67	62.58
1972	32.91	83.88	11.28	2.74	32.96	62.64
1973	33.16	84.10	11.26	2.74	33.08	62.52
1974	33.41	84.28	11.28	2.75	33.31	62.54
1975	33.66	84.46	11.30	2.73	33.52	62.54
1976	33.91	84.65	11.33	2.72	33.71	62.52
1977	34.17	84.83	11.35	2.71	33.90	62.52
1978	34.42	85.01	11.37	2.71	34.05	62.48
1979	34.67	85.19	11.39	2.70	34.27	62.52
1980	34.92	85.38	11.42	2.69	34.40	62.48
1981	35.17	85.56	11.44	2.68	34.49	62.41
1982	35.42	85.74	11.46	2.68	34.71	62.47
1983	35.68	85.92	11.48	2.67	34.87	62.47
1984	35.93	86.11	11.51	2.66	35.05	62.51
1985	36.18	86.29	11.53	2.66	35.14	62.46
1986	36.43	86.47	11.55	2.65	35.30	62.48
1987	36.70	86.72	11.58	2.69	35.34	62.39
1988	36.96	86.97	11.61	2.72	35.54	62.47
1989	37.23	87.23	11.64	2.76	35.55	62.37
1990	37.50	87.48	11.67	2.80	35.70	62.4
1991	37.77	87.73	11.70	2.83	35.86	62.46
1992	38.03	87.98	11.74	2.87	35.98	62.47
1993	38.30	88.24	11.77	2.91	36.15	62.55
1994	38.57	88.49	11.80	2.95	36.14	62.45
1995	38.83	88.74	11.83	2.98	36.09	62.30
1996	39.10	88.99	11.86	3.02	36.31	62.43
1997	39.37	89.25	11.89	3.06	36.27	62.32
1998	39.64	89.50	11.92	3.10	36.44	62.41
1999	39.90	89.75	11.95	3.13	36.55	62.45
2000	40.17	90.00	11.98	3.17	36.58	62.41
2001	40.44	90.25	12.01	3.21	36.62	62.45
2002	40.70	90.51	12.04	3.24	36.60	62.42
2003	40.97	90.76	12.07	3.28	36.56	62.38
2004	41.24	91.01	12.10	3.32	36.58	62.41
2005	41.51	91.26	12.13	3.36	36.58	62.41
2006	41.74	91.52	12.14	3.41	36.52	62.34
2007	41.98	91.77	12.16	3.47	36.61	62.43

Year	Electrolysis (%)	Mechanical drive (%)	Cooling (%)	Light (%)	Heat (%)	HTH (aluminum) (%)
2008	42.22	92.02	12.17	3.52	36.61	62.43
2009	42.46	92.28	12.18	3.58	36.50	62.33
2010	42.69	92.53	12.19	3.63	36.56	62.38
2011	42.93	92.79	12.20	3.69	36.49	62.32
2012	43.17	93.04	12.21	3.75	36.58	62.41
2013	43.41	93.29	12.22	3.80	36.61	62.43
2014	43.64	93.55	12.23	3.86	36.57	62.39

Source: Felício et al. [15].

Table A1.
Industry exergy efficiencies by type of uses.

Appendix B

See **Table A2.**

Year	LTH2 (%)	Light (%)	Mechanical drive (%)	Cooling (refrig) (%)	Cooling (AC) (%)	Electronics (%)
1970	20.18	0.90	76.33	14.70	4.17	0.04
1971	20.32	1.14	76.41	14.68	4.17	0.04
1972	20.46	2.74	76.50	14.66	4.17	0.04
1973	20.21	2.74	76.59	14.64	4.17	0.04
1974	20.24	2.75	76.72	14.67	4.17	0.05
1975	20.24	2.73	76.84	14.69	4.17	0.06
1976	20.21	2.72	76.97	14.72	4.17	0.08
1977	20.21	2.71	77.09	14.75	4.17	0.09
1978	20.13	2.71	77.22	14.77	4.17	0.10
1979	20.21	2.70	77.34	14.80	4.17	0.11
1980	20.13	2.69	77.47	14.82	4.17	0.13
1981	19.9	2.68	77.59	14.85	4.17	0.14
1982	20.1	2.68	77.72	14.88	4.17	0.15
1983	20.10	2.67	77.84	14.90	4.17	0.16
1984	20.18	2.66	77.97	14.93	4.17	0.18
1985	20.07	2.66	78.09	14.96	4.17	0.19
1986	20.13	2.65	78.22	14.98	4.17	0.20
1987	19.94	2.69	78.31	15.02	4.24	0.27
1988	20.10	2.72	78.40	15.06	4.31	0.34
1989	19.88	2.76	78.49	15.09	4.38	0.41
1990	19.96	2.80	78.59	15.13	4.45	0.49
1991	20.07	2.83	78.68	15.16	4.52	0.56

Year	LTH2 (%)	Light (%)	Mechanical drive (%)	Cooling (refrig) (%)	Cooling (AC) (%)	Electronics (%)
1992	20.10	2.87	78.77	15.20	4.59	0.63
1993	20.27	2.91	78.87	15.24	4.66	0.70
1994	20.05	2.95	78.96	15.27	4.73	0.77
1995	19.74	2.98	79.05	15.31	4.80	0.84
1996	20.02	3.02	79.14	15.34	4.87	0.92
1997	19.77	3.06	79.24	15.38	4.94	0.99
1998	19.96	3.10	79.33	15.42	5.01	1.06
1999	20.05	3.13	79.42	15.45	5.08	1.13
2000	19.96	3.17	79.52	15.49	5.15	1.20
2001	20.05	3.21	79.61	15.53	5.22	1.27
2002	19.99	3.24	79.70	15.56	5.29	1.34
2003	19.91	3.28	79.80	15.60	5.36	1.42
2004	19.96	3.32	79.89	15.63	5.43	1.49
2005	19.96	3.36	79.98	15.67	5.50	1.56
2006	19.83	3.41	80.09	15.68	5.16	1.60
2007	20.02	3.47	80.21	15.70	4.83	1.64
2008	20.02	3.52	80.32	15.71	4.50	1.69
2009	19.80	3.58	80.43	15.72	4.16	1.73
2010	19.91	3.63	80.54	15.73	3.83	1.77
2011	19.77	3.69	80.66	15.75	3.49	1.81
2012	19.96	3.75	80.77	15.76	3.16	1.86
2013	20.02	3.80	80.88	15.77	2.82	1.90
2014	19.94	3.86	80.99	15.78	2.49	1.94

Source: Felício et al. [15].

Table A2.
Commerce and services exergy efficiencies by type of uses.

Author details

Teles Huo[1*], Miguel St. Aubyn[2], Laura Felício[3] and Tânia Sousa[3]

1 Faculty of Economics, Eduardo Mondlane University (UEM), and Regional Center for Excellence in Oil and Gas Engineering and Technology Studies – UEM (CSOGET), Maputo, Mozambique

2 Research Unit on Complexity and Economics (UECE), Higher Institute of Economics and Management (ISEG), Lisbon University, Lisboa, Portugal

3 Marine Environment and Technology Center (MARETEC), Institute for Systems and Robotics (LARSyS), Higher Technical Institute (IST), Lisbon University, Lisboa, Portugal

*Address all correspondence to: teles.huo@uem.mz

References

[1] Kümmel R. The Second Law of Economics: Energy, Entropy, and the Origins of Wealth. London: Spinger; 2011. DOI: 10.1016/j.strueco.2013.07.004

[2] Serrenho AC, Warr B, Sousa T, Ayres RU, Domingos T. Structure and dynamics of useful work along the agricultural-industry-services transition: Portugal from 1856 to 2009. Structural Change and Economic Dynamics. 2015;**36**(C):1-21. DOI: 10.1016/j.strueco.2015.10.004

[3] Santos J, Domingos T, Sousa T, Aubyn M. Useful exergy is a key in obtaining plausible aggregate production functions and recognizing the role of energy in economic growth: Portugal 1960-2009. Ecological Economics. 2018;**148**(C):103-120. DOI: 10.1016/j.ecolecon.2018.01.008

[4] Oladiran MT, Meyer JP. Energy and exergy analyses of energy consumptions in the industrial sector in South Africa. Applied Energy. 2007;**84**(10):1056-1067

[5] Heun MK, Brockway PE. Meeting 2030 primary energy and economic growth goals: Mission impossible? Applied Energy, Elsevier. 2019;**251**(C):1-1. DOI: 10.1016/j.apenergy.2019.01.255

[6] Wall G. Exergy tools. Part A: Journal of Power and Energy. 2003;**217**(2):125-136

[7] Ayres RU, Ayres LW, Pokrovsky V. On the efficiency of US electricity usage since 1900. Energy. 2005;**30**:1092-1145

[8] Pinto R, Henriques ST, Brockway PE, Heun MK, Sousa T. The rise and stall of world electricity efficiency: 1900-2017, results and insights for the renewables transition. Energy. 2023;**269**(C):1-14. DOI: 10.1016/.J.ENERGY.2023.126775

[9] Karakatsanis G. Exergy and the economic process. Energy Procedia. 2016;**97**:51-58. DOI: 10.1016/j.egypro.2016.10.018

[10] Wall G. Exergy: A useful concept within resource accounting. Report no. 77-42, Institute of Theoretical Physics, Chalmers University of Technology and University of Göteborg, S-412 96 Göteborg, Sweden, 1977; p. 1-61

[11] Dos Muchangos A. Moçambique paisagens e regiões naturais. Maputo: Tipografia Globo; 1999

[12] Brockway PE, Sorrell S, Foxon T, Miller J. Exergy economics: New insights into energy consumption and economic growth. In: Jenkins KEH, Hopkins D, editors. Transitions in Energy Efficiency and Demand. London: Routledge; 2018. pp. 133-155

[13] Ayres RU, Van Den Bergh JCJM, Lindenberger D, Warr B. The underestimated contribution of energy to economic growth. Structural Change and Economic Dynamics. 2013;**27**(C):79-88

[14] Sousa T, Brockway PE, Cullen JM, Henriques ST, Miller J, Serrenho AC, et al. The need for robust, consistent methods in societal exergy accounting. Ecological Economics. 2017;**141**:11-21. DOI: 10.1016/j.ecolecon.2017.05.020

[15] Felício L, Henriques ST, Serrenho A, Domingos T, Sousa T. Insights from past trends in exergy efficiency and carbon intensity of electricity: Portugal, 1900-2014. Energies. 2019;**12**(3):1-22. DOI: 10.3390/en12030534

[16] Serrenho AC, Sousa T, Warr B, Ayres RU, Domingos T. Decomposition of useful work intensity: The EU (European Union)-15 countries from 1960 to 2009. Energy;**76**(C):704-715. DOI: 10.1016/j.energy.2014.08.068

[17] República de Moçambique, Ministério da Indústria e Comércio. Política e Estratégia Industrial 2016-2025, Maio de 2016. 2016

Chapter 2

Perspective Chapter: The Exergetic Friction Fluctuations at the Boundary Lubrication

Rodrigo Bulnes A.

Abstract

Since friction represents the core of the historical analysis of the industrialization process of the economy, it emerges as pertinent to analyze the conservation of energy, its entropic dissipation through friction in particular and the correlation of physical forces, which play a crucial role in the analysis "Lubricated Machinery-Industry on a large scale." In this sense, the scattered pseudo-lubricated boundary friction experimental results (typical lubrication regime of the large mining machinery) are reordered with exergetically dissipated friction theory to demonstrate that mechano-chemical reaction kinetics between surfaces emerge as an innovative third relative (t/τ)-time coordinate that helps to eliminate this characteristic scatter of such experimental data. These findings suggest furthering the concept of Friction-e (Economy, Energy, and Emissions).

Keywords: friction, fluctuations, dissipated, time, exergy and environment

1. Introduction

Historically, friction represents the nucleus of the analysis of industrial development, an aspect thus considered in various studies that show the important economic losses due to friction and wear, especially in the world mining [1].

The important total energy consumption of mining is mainly used to overcome friction and combat wear, which in turn generate CO_2 emissions; that is to say, it is possible to propose a concept of Friction-e (Economy, Energy, and Emissions) to refer to the interface of friction in the ecology of mining metabolisms and their derivatives: heat and wear.

From the Tribology, the circular economy exceeds the life cycle, the maintenance, the physical asset management and the supply chain. Additionally, it incorporates the socio-ecological dimension defined in the "Eco-Tribology," meaning that it is necessary to "reduce the complexity" of mechanical-chemical reactions at the level of the friction interfaces. In this sense, the Eco-Tribology is a nucleus of innovations and technological changes that allow to impact on the economy of friction and wear, pollution control, reduction of emissions and increased conservation and efficiency of energy. In South America (Chile, "mining" country), the mining requires better socioecological standards. The policies of sustainability, circular economy, greenhouse gas emissions and climate change that arise, find in the Friction-e a possibility to inform and act on the socioeconomic metabolism for sustainable mining.

2. Friction at the interface of the economy, energy and emissions metabolisms of industrial mining

2.1 On friction during the exergetic exhaustion of minerals in mining

Satisfying the global demand for mineral resources for the renewable energy and electromobility industries represents a complex environmental challenge where the demand for the "new oil" (Copper) directly pressures the socioenvironmental standards and regulations required in the South American Andean region.

In this way, the Chilean mining strategy has defined a "road map" focused on promoting the digitization of its processes, the use of renewable energies, the development of "green hydrogen," and energy efficiency as guiding principles of a national policy of decarbonization (it is what could be called "Copper to produce still more Copper").

However, environmentally speaking, Copper is increasingly difficult to obtain: as mining progresses, ore grades tend to drop, the material becomes harder to grind, more waste is generated, more water and energy are consumed per pound produced, the mines are getting deeper and, in ecologically sensitive areas—decades after the closure of the mines—the tailings still represent a serious environmental problem, etc. In simple terms, deposits are ecologically complex to sustain, with the mining industry opting to replace direct labor by digitizing it most of the time. Thus, and as a more or less (in) direct recognition that the environmental-mining dilemma has a genuinely scientific basis, instead of "green Copper" it is preferred to migrate toward "traceable Copper."

Summarizing, the regulations imply the socially symbolic, the politically accepted in this context of "eco-sustainable" production of minerals, with the material and energetic shielding being the hardest and most factual part that provides said eco-production as tribological metabolisms of dissipative structures that create or destroy the exergetic reserve of minerals during their transformation (exploitation) [2].

On the one hand, before it can be used, the exergy must be transformed into a quantifiable process called useful work. On the other hand, and unlike energy, exergy is not conserved, it is "used up" and partially degraded as the appearance of dissipative structures during the friction that implies the production of minerals [3]. That is, the dissipated useful work is exergetically exhausted until the disappearance of said structures or, in other words, a system dissipates useful work that would be tribologically less with respect to the available exergy input to it.

During the transformation of exergy into useful work, it will be the thermodynamic losses and, therefore, the dissipative structures that will succumb as a consequence of the Second Law of Thermodynamics; that is, the fraction wasted during production will be dissipated in the generation of irreversible entropy, where the fraction of exergy that is lost depends on the efficiency of the transformation process, and on the technology and science used in it. Hence the importance of incorporating "Tribological Science" into the socioeconomic metabolism of mining, since it acts informing from "friction" as a historically fundamental factor for large-scale industries [4].

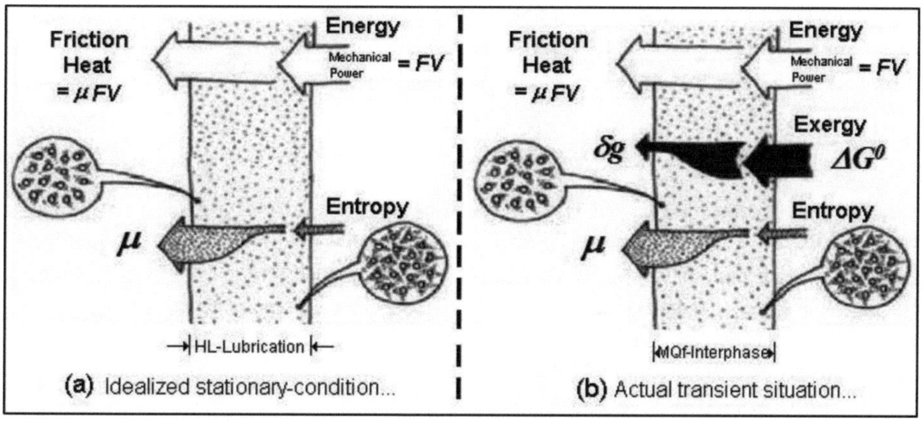

Figure 1.
Appealing to an idealized stationary friction condition at the Hydrodynamic Lubrication-HL (a), obviously does not explain the appearance of wear, etc. In operational practice (b), it is known that friction (μ) becomes unstable (transient); in fact, the Lubricant-EHL film collapses 75% of the time, which would imply that the tribo-interface dissipates a Friction-MQf of exergetic origin as responsible for wear. F: *force and* V: *velocity.*

The following **Figure 1** represents the difference between energy and friction exergy in the "depletion" process of the structures of a generic tribo-interface between bodies.

Summarizing, the artificial is deterministic and reversible (digital technology), while the natural contains essential elements of chance and irreversibility; for this reason, "Non-Equilibrium Irreversible Thermodynamics" calls for a new vision of matter in which matter is no longer passive (a purely mechanical concept), but rather an associated spontaneous activity (dissipative structures) emerges. This change is so profound that for almost four decades, there has been talk of a new dialog between man and nature [5].

2.1.1 The exergetic dissipation of friction in the metabolism of mining

As a factor of metabolism that creates/destroys the exergy reserve of minerals, friction represents a fissure between matter and energy dissipated by friction of a mechanochemical nature (abbreviated Friction-MQf) [6]. In this way, reducing that crack to the complexity of the mechano-chemical reactions at the Friction-MQf interface level encourages us to understand what is a friction interface?, a pertinent question since the effort made by Tribology consists in the development of wear-resistant materials that help reduce socioeconomic losses by 40% over the next 10 to 15 years [7].

2.2 What is a friction interface? (contribution of non-equilibrium irreversible thermodynamics: Friction-MQf)

Simultaneously dissipating heat from purely mechanical friction and exergy of a mechanochemical nature, the friction interface is a third tribological body where one of the surfaces corresponds to a first body, the same as the other to a second body [8]. So, the third tribological body has its own composition materially dissipated in an exergetic potential that is more or less responsible for the heat of friction and wear. In simple terms, during the transmission of mechanical power, the generation of frictional heat and wear are affected by the times of kinetics of physicochemical reactions coming from the third friction interface body that dispels them.

The following simple expression summarizes these concepts:

$$\dot{\mu} + \phi(t/\tau)\mu - \beta\mu^2 \sim 0 \tag{1}$$

where μ is Friction-MQf, β a stationary thermomechanical factor of the materials, and $\varphi(t/\tau)$ a propagator-operator linked to the *in situ* kinetics (t/τ) as referring to the dissipated exergy (literally speaking, Eq. (1) considers the tribo-contact between bodies as an intense physical-chemical reactor capable of dissipating a relative time coming from an intense mechanochemical reaction until today misnamed "coefficient" of purely mechanical friction).

The deduction of this Eq. (1) develops as follows:

Is well known that in pure exothermic adsorption states $(\varphi[\Delta g_{ads}])$, the friction coefficient monotonically decreases along with the sliding speed at the boundary lubrication EP condition [8]. For instance, this results in the well-known increase of the load-carrying capacity as a consequence of the adsorption of the molecules of additive on the rubbing surfaces (see [9, 10]). In this case, the monotonous decrease vs. velocity of μ, "attract" the instabilities that would cause the emergence of an alternative and spontaneous nonequilibrium thermodynamic regime into the frictional process, in a way that would look as the thermodynamic fluctuations of the quantities $du, dv, ds, \ldots$ do not perform any role in the dynamics of μ. Let us agree on, then, that our tribo-system shows a homogeneous exothermic adsorption state, so that queue this purely adsorptive phenomenon can be expressed—with no effective influence from fluctuations—as a regressive relationship between the friction coefficient and the sliding speed:

$$\mu \sim 1/V. \tag{2}$$

By developing this expression with $\mu \cong \frac{d\mu}{d\left(\frac{1}{V}\right)} \cdot 1/V$, where $d\mu/d\left(\frac{1}{V}\right)$: gradient, Eq. (2) corresponds to:

$$\begin{aligned} d\mu &\cong d(1/V) \cdot V \cdot \mu \quad / * (dt)^{-1} \\ \dot{\mu} &\cong -V^{-2} \cdot \left(\frac{dV}{dt}\right) \cdot V \cdot \mu \\ \dot{\mu} &\cong -\left(\frac{a}{V}\right) \cdot \mu \end{aligned} \tag{3}$$

with a: acceleration (see "*A commentary*," more below). A very particular result for the equation above is (see [11]):

$$\dot{\mu} \cong -\omega_V \cdot \mu \Rightarrow \frac{\dot{\mu}}{\mu} \cong -\omega_V, \text{ (pure ads.} = \phi[\Delta g_{ads}]) \tag{4}$$

with ω_V *[Hz]*: denoting contact vibrations or "exciting mechanical macro-force" on the friction process (but we do not forget that the μ-process is intimately related to the Gibbs' free energy).

If one still considers an increase of the load capacity but this time as a *degressive relationship* between the friction coefficient and the sliding speed, one would consider that the pure adsorption state is subject to a competitive environment; in other words, it is affected by a perturbation toward the state of desorption (of course this represents a fluctuation of the initial activity regime from $\varphi[\Delta g_{ads}]$ to states $\varphi[\Delta g_{ads} - \Delta g_{des}]$). In mathematical terms, the degressive—and must be appointed—quasi-expontaneous with velocity effect can be *degressively* expressed as:

$$\mu \sim \left(\mu_{eq.} - b^V\right) \tag{5}$$

where $\mu_{eq.}$ indicates, now, the unstable adsorptive equilibrium condition, and b is a number. By calculating the derivative of the equation above: $\dot{\mu} \sim -a \cdot (log_e b) \cdot b^V = a \cdot (log_e b) \cdot \Delta\mu$, with $\Delta\mu \sim \left(\mu - \mu_{eq.}\right) \neq 0$, Eq. (5) result can be expressed in terms of a series of powers around $\Delta\mu$:

$$\dot{\mu} \sim C_1 \cdot a \cdot \left[\delta\mu + (1/2) \cdot \delta^2\mu + \ldots\right], \tag{6}$$

where C_1 indicates a constant.

Evidently, the first-order $\delta\mu$ fluctuation refers to Eq. (4). On the other hand, and to give physical sense to the second-order $\delta^2\mu$ fluctuation, we began to accept that the problematic of the friction refers to the own considerations of a dissipated and irreversible process. That is to say, it is valid that the "entropic evolution (S)" of frictional heat can acquire the following formulism: $Q(\mu)/\theta = S \Rightarrow \Delta\mu \sim \left(S - S_{eq.}\right) = \delta S + \frac{1}{2} \cdot \delta^2 S + \ldots$, where Q indicates frictional heat a θ temperature. But our interest is focused on a situation at irreversible nonequilibrium where the entropy is not necessarily at the maximum (the friction system is dissipative or not a closed system). That is to say [12], the imposition of a gradient on μ (e.g. temperature, velocity, etc.) dissipates $Q(\mu)$-energy associated with qualitative changes in the entropy rate in the friction system. These qualitative changes are directly influenced (affected) by, at least, second-order terms δ^2 of the entropy ([13], pp. 301–308). Thus, we accept second-order terms for the change in the dynamics of μ because now the frictional system—which is sufficiently unstable, etc.—is not at total adsorptive equilibrium (as aforementioned, there is desorptive competition from $\phi[\Delta g_{ads}]$ to $\phi[\Delta g_{ads} - \Delta g_{des}]$ into the friction system). From the Eq. (6):

$$\dot{\mu} \cong C_2 \cdot a \cdot \mu^2 \tag{7}$$

where $\mu \sim \delta\mu \Rightarrow \delta\mu^2 \sim \mu^2$, $C_2 \cdot a \cong \frac{a}{\gamma \cdot \theta}\left[hz^2\right]$ and γ $[m/m_0 K]$: coefficient of thermal expansion of the material (the thermodynamic nature of γ allows the association, via θ, with a nonequilibrium situation. This choice negatively affects the use of a length unit $[m]$ obtained traditionally from Hertz's static homogeneous deformation at equilibrium.

Joining together Eqs. (4) and (7) in one single expression (plus additives constant), we have the initial Eq. (1):

$$\dot{\mu} + \omega_V \mu - \beta\mu^2 = 0 \tag{8}$$

where $\beta = \frac{a}{\gamma \cdot \theta}$.

The solution to nonlinear Eq. (8) can be formulated with two methods. The first method provides a "closed" solution with β as a parametric coefficient. The second method is iterative and is more "realistic" physically, since it lets one consider the fluctuations of a thermo-mechanical quantity of the material, that is to say, β:

$$\dot{\mu} + \omega_V \mu - \beta \mu^2 = 0 \begin{cases} \mu(t/\tau) = \dfrac{\mu_0}{e^{\omega_V t} + \zeta \cdot \mu_0 \cdot (1 - e^{\omega_V t})} \\ \mu_{t+1} = -\omega_V \cdot \mu_t \cdot (1 - \zeta \cdot \mu_t) \end{cases}, \quad (9)$$

with $\zeta = \left(\frac{\beta}{\omega_V}\right)$: a damping factor of thermo-mechanical nature. But: $\frac{\beta}{\omega_V} = \frac{a}{\gamma \cdot \theta} \cdot \frac{1}{\omega_V} \simeq \frac{V}{\gamma \cdot \theta} \Rightarrow \zeta \cong \frac{P_e}{\gamma \cdot \theta}$, where $P_e(V, l, \rho, C_P, \lambda)$ is Peclet's number which contains, obviously, material parameters.

Both solutions above give the indication that the irregular trajectory of the friction coefficient in time is the result of the fluctuations of the thermo-mechanical quantities of the material bodies ($\delta\gamma$, $\delta\lambda$, $\delta\rho$, $\boldsymbol{\delta C_P}$, $\delta\lambda$). On the one hand, the first closed solution immediately indicates that the tribological failures (scuffing, pitting, micro-pitting, etc.) may be interpreted—in coherence with ω_V—as "thermo-mechanical resonance" phenomena (see **Figures 2–4**).

The second iterative solution in Eq. (9) indicates that the origin of the trajectory of the friction coefficient in time is explicitly chaotic (logic map).

With these preliminary results in mind, one may suggest that the present theory manages to physically explain—under minimum and plausible hypotheses—concepts ranging from the fluctuation of the friction coefficient to the *in situ* load carrying capacity at the scuffing, pitting, micro-pitting, etc., limits. Simply, the practical application of this new thermodynamic definition for friction suggest that modern Nonlinear Thermodynamics is the correct method to approach, in a simple and direct way, some tribological

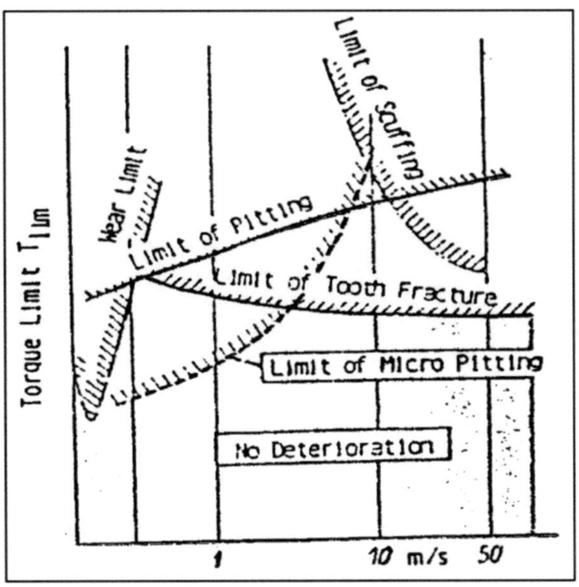

Figure 2.
Load carrying capacity (lubricant-related failures in gears, oil film collapse, additive depletion, and use of improper lubricant for the application).

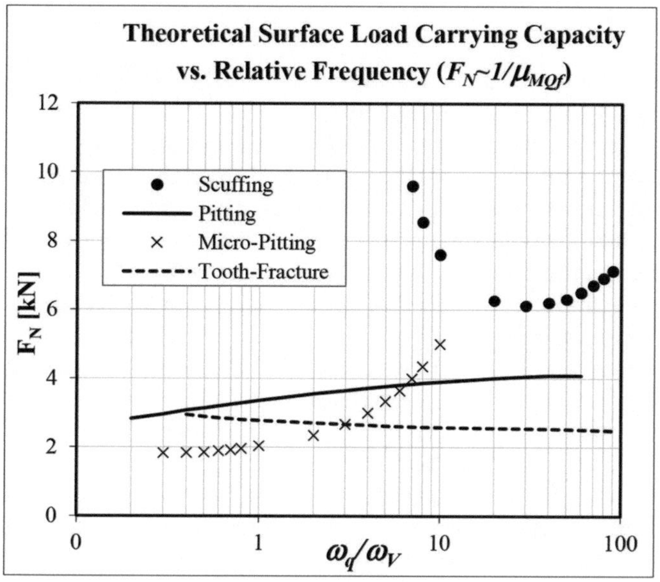

Figure 3.
Load carrying capacity (lubricant-related failures in gears, oil film collapse, additive depletion, and use of improper lubricant for the application), according to friction-MQf.

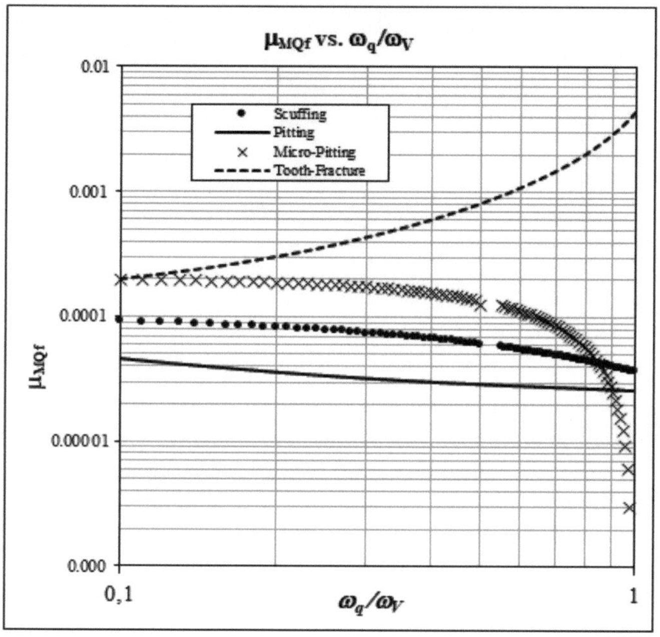

Figure 4.
The exergetic friction-MQf (competition between purely mechanical vibrations vs. those dissipated by the tribo-chemical interface).

problems of the metal-mechanical industry, that is, the nonequilibrium thermodynamic definition from the friction coefficient makes unnecessary the use of any Norme to calculate *in situ* the load-carrying capacity of a friction couple.

2.3 A commentary

It is evident that a rapid identification of "a" equal to dV/dt (see Eq. (3)), forces us to make equal the micro-dynamic aspects a from "mechanical friction movement," with the purely macro kinetics V ($a = dV/dt$ it is a fundamentally mechanical perspective that do not explore the reasons why the forces that accelerate the "movement" of the interface process). From this perspective, extremely purist and probably inadequate for tribological problems, the ω_V mechanical contact vibration would be the cause and the origin of all the phenomena observed in relation to the friction problematic. Nevertheless, the fact that appears a kind of acceleration forces us to wonder if the act of deriving the dV sliding velocity in relation to the time dt (what time?), is it not equivalent to being involved with the time scales from the frictional phenomenon itself? because it is an empirical fact that despite of the $V = cte.$ ($a = 0$), the process μ fluctuates at nonequilibrium. In fact, the act of involving with the dt represents the delicate action of "scaling the friction phenomenon" throughout of its own time scale, due to the necessary appearance of an internal time τ. Thus, the forces that accelerate the "relative movement" of the interfacial process are not in the plane $dt \sim \omega_V$ from the contact vibrations, in which you could say, in general terms, that the a-acceleration must be identified with something more fundamental like the stimulation of the interfacial chemical reactions -or mechanochemistry activation—due to the friction, but a scale $d(t/\tau) \sim (\omega_q/\omega V)$ (this topic is handled precisely by the Tribochemistry [14]).

Then, we agree on the fact that last Eqs. (3) suggests a mixture between the Tribochemistry (represented by $a = a(t,\tau)$) and the kinetics from the sliding movement (represented by the contact vibrations ω_V provided by $V=V(t)$). As in both aspects we do not have, necessarily, to share identical scale probabilistic of time; thus, it is understandable that a (or the Tribochemistry) and V (or the contact vibrations) convolution with each other, via delay τ-time like denoting the rigorous character probabilistic of all experimental or theoretical measurement. In other words, exists an intrinsic stochastic aspect in the combined variable $a/V \sim \varphi(t/\tau)$, in a way that Eq. (9) must be correct by the most general expression:

$$\dot{\mu} + \left(\frac{\omega_q}{\omega_V}\right) \cdot \phi(t/\tau)\mu - \beta\mu^2 = 0, \tag{10}$$

where ω_q is a natural frequency with a mechanochemistry origin, and φ denotes a stochastic propagator or a function of probabilities. Eq. (10) justifies again the assumption about the underlying dynamic that provokes the frictional phenomenon fluctuations of μ (the so called frictional dynamic does not have to be take it for granted that always as reducible a white noise (or φ = normal distribution), which is undoubted, reversible or with no memory).

The denomination of "coefficient" of friction (Classical Laws of Tribology) is inadequate due to the mere presence of this dissipated physical-chemical relativistic time (t/τ). This implies that the "wear" effect is not proportional to the cause "heat"; that is, and from the theoretical and idealized purely mechanical friction of heat, the transmission of work and energy *via* (t/τ), suggests that the tribo-interface "accommodates" in speed (the friction interface would possess its own relativistic mechanics, its own relativistic (t/τ)-time dissipating it to its immediate surroundings during the process of accommodation of tribo-mutations between surfaces).

Summarizing, the friction heat is associated with a nominally calculated friction from the idealized purely mechanical friction force (heat), but the wear would rather be related to the simultaneous dissipation of a useful energy, an exergetic deviation hitherto "hidden" in that thermodynamic nonpotential that is the heat of friction: the friction interface dissipates matter, energy, relativistic time and exergy simultaneously.

For example, according to the Classical Laws of Tribology, friction is independent of speed, something that a Friction-MQf with (t/τ) refutes mathematically, chemically, and physically: during thermodynamic dissipation, which supposes the ideal transmission of "tribological power" (velocity-independent friction), curiously implies that, on the contrary, the friction exergy refers to the practical reality of a loss of useful work of the friction power in terms of eventual wear ..., not only being the heat of friction responsible for it (the responsible potential would be the exergy of friction that provides a measure of the effort that nature makes to replace and/or return to it the substances that make up the material of the bodies, wearing them down or not).[1]

Being this exergetic dissipation a measure of the deviation of the state of a thermodynamically real system with respect to the immediate ideal environment (mechanical notion of the force of friction), the exergetic dissipation of the friction interface results in an attribute of the "tribological system" and their environment together. We call this program "Mechanochemical Friction-MQf," and we identify in the Gibbs Free Energy the thermodynamic potential dissipated by the friction exergy that wears the materials (Friction-MQf has units $W/kg°K^2$). Said program points out that together with the generation of friction heat, simultaneously in situ an exergetic potential is dissipated that may or may not wear out the surfaces. If the potential does not wear, then it is similar to STP (zero-wear), but if it does wear, then the "sensitive friction" (mechanics) is influenced by a relative time trajectory that can be calculated. Consequently, said (t/τ)-time can inform about its effects on the Mining-Industrial eco-sustainability in terms of exergetic depletion of mineral resources (Friction-e).

Figure 5 explains the value contribution of the exergy dissipating in friction-MQf interface:

Coming from the interface-MQf, the relative time (t/τ) is born, which surprisingly manages to "order" the messy measurements of tribological tests that usually consider friction vs. speed, vs. contact force, vs. temperature or another measurement of the "conventional mechanistic tribology." To deepen this concept of Friction-MQf, **Figure 5** shows the tabulation "disorderly of purely mechanical friction" vs. speed at different temperatures obtained for 02 types of nonlubricated ceramic materials, comparing it with the "correction" suggested by the relative time dissipation (t/τ) proposed by Friction-MQf.[2]

[1] In Tribology there is the phenomenon of "zero-wear" as dissipation that arises from the "selective transfer phenomenon-STP", that is, it is a type of "circular wear" where matter and energy circulate between surfaces, without leaving them (wear tends to zero). The "thermodynamic engine" of the STP implies an exergy potential, and not frictional heat (which is not a thermodynamic potential).

[2] R. Bulnes, TU-Dresden 2001 (this conventional tribological data was extracted from more or less confidential measurements published during the year 2000, "Studie zur Untersuchung von Minimalmengenschmierung und Trockenlauf für Getriebe," Forschungsvorhaben Nr. 366 I + II + III, Vol. 613 de Forschungsheft: Forschungsvereinigung Antriebstechnik).

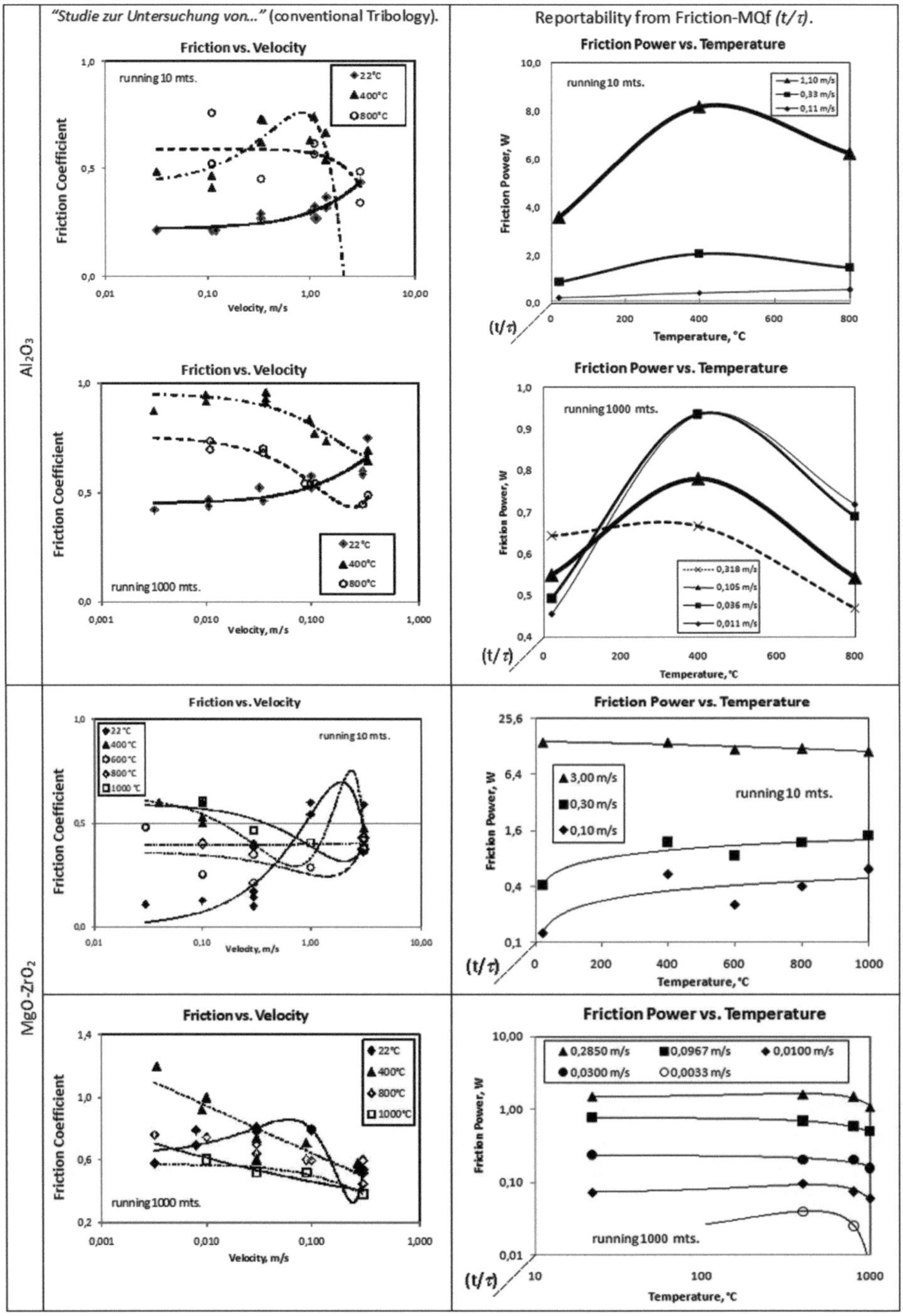

Figure 5.
The scattered friction experimental results (left), are reordered with exergetically dissipated Friction-MQf theory (right). That is to say, the mechanochemical reaction kinetics between surfaces emerge as an innovative third relative (t/τ)-time coordinate that helps to eliminate the characteristic scatter of the conventional tribological experiments.

3. Conclusions

Highly coherent with the Godet's third body notion, the present theory about the Friction-MQf coefficient μ, allows to access—in simply and direct form, and with a minimal hypotheses—to the *in situ* calculation of the load carrying capacity at the enigmatic scuffing, pitting, micro-pitting and tooth-fracture limit damages (the Physics of the load carrying capacity is an issue of dissipated exergy and does not require any Norm in order to assess it). The theoretical results like this obtained are especially susceptible to be considered when lubrication failures appear.

On the other hand, the idea of linking the friction process with the principles of the Nonlinear Thermodynamics allows to overcome the theoretical contradiction, which means to simulate the irreversible friction process *via* the reversible second Newton's Law, i.e., *via* the standard concept of friction force. Summarizing, the entropy rate of the friction process identifies to μ as a thermodynamic reaction sensitive to Gibbs' Free Energy activity regime of the friction process. From this point of view, the frictional temperature only "triggers" levels of Free Energy which the friction coefficient uses to modify its trajectory in the relive (t/τ)-time, that is, the irreversible δ^2, δ^3, fluctuations of μ would be a measure of the physical and chemical activity ensemble taking place on the actual contact area. This situation matches the idea that there is no direct and defined relationship between the friction temperature and the friction coefficient. On this matter, although the friction coefficient varies whith the sliding velocity, normal load, composition of the surfaces, etc., such a variation would be just a functional of the Free Energy developed by the exergetical friction-MQf.

To sum up, the variations of the friction coefficient are due to the internal dissipation of the physical-chemical regime, and not due to the external dissipation of frictional heat $(\overline{\mu}\cdot F_N V)$. Therefore, that thing which takes over the path of μ in (t/τ)-time is the Gibbs' Free Potential and is not the dissipated energy in way that something which is not a thermodynamic potential, that is, in a way of frictional heating. This simple conclusion could take to very important consequences in the form to approach the Physics of the tribo-contact, because the frictional heating has been, from always, the axiomatic final results of all tribo-energetical balance which is "supposedly correlated" with μ. This assumption—which is explicit here questioned—is something that the tribologist, in general, has trivially accepted until today.

Nomenclature

C_P	specific heat [J/kgK];
g	Gibbs' free potential (energy) [J/kg];
l	length of static deformation from Hertz [m];
F_N	load-carrying capacity [N]
V	speed [m/s];
θ	stationary temperature [K]
$\mu\ (t/\tau)$	exergetical friction-MQf coefficient (Dissipation)
μ_O	referential friction (mechanical) coefficient (state)
t	time [s]
τ	delay time [s]
t/τ:	normalized time [−]
φ	denotes function;

γ	coefficient of thermal expansion [m/m$_0$K]
ρ	density [kg/m^3]
λ	thermal conductivity [W/mK]
ω_V	contact mechanical vibrations [hz]
ω_q	natural frequency of mechanochemistry origin [rate]

Author details

Rodrigo Bulnes A.
Tribothek Sytems SpA, Santiago, Chile

*Address all correspondence to: rbulnes@gmail.com

References

[1] Holmer K et al. Global energy consumption due to friction and wear in the mining industry. Tribology Interna tional. 2017;**115**:116-139

[2] Valero A, Valero A. Accounting for mineral depletion under the UN-SEEA framework. 2018. DOI: 10.5772/intechopen.77290

[3] Ayres R, Warr B. The economic growth engine: How energy and work drive material prosperity. 2009. Available from: https://www.researchga te.net/publication/285923899

[4] Burkett P, Bellamy J. Metabolism, energy, and entropy in Marx's critique of political economy: Beyond the Podolinsky myth. Theory and Society. 2006;**35**:109-156

[5] Prigogine I. Only an Illusion, The Tanner Lectures on Human Values. Clare Hall: Cambridge University editor's, Jawaharlal Nehru University; 18 Dec 1982

[6] Bulnes R. Mechanochemical friction of third-body as an exergetic collision. Tribology On-Line (TROL). 2011;**6**(1):55-63 Japanese Society of Tribologists

[7] Zhai W et al. Recent Progress on wear-resistant materials: designs, properties, and applications. Advanced Science. 2021;**8**:2003739

[8] Godet M. The third-body approach: A mechanical view of wear. Wear. 1984; **100**:437-452

[9] Jahanmir S, Beltzer M. Effect of additive molecular structure on friction coefficient and adsorption. ASME Journal of Tribology. 1986;**108**: 109-116

[10] Cameron A. On a unified theory of boundary lubrication. In: Proceedings of the 11th Leeds-Lyon Symposium on Tribology. Butterworths; 1984. pp. 94-99

[11] Escobar E. The additive EP condition and the critical scuffing limit for rolling-sliding. ASME: Journal of Tribology. 1996;**118**:125-130

[12] Critical Problems in Physics. In: Fitch V, Marlow D, Dementi M, editors. Princeton Series in Physics. 1997. p. 51

[13] Kondepudi D, Prigogine I. Modern Thermodynamics: From Heat Engines to Dissipative Structures. New York: John Wiley & Sons; 1998

[14] Muratov V, Luangvaranunt T, Fischer E. The tribochemistry of silicon nitride: effects of friction, temperature and sliding velocity. Tribology International. 1999;**31**(10):601-611

Section 2

Exergy Applied to Energy

Chapter 3

Comparative Exergy Analysis of Thermal Power Plant Systems

Pasupuleti Ravindra Kumar and Naradasu Ravi Kumar

Abstract

The concept of exergy, derived from the second law of thermodynamics, becomes a valuable source tool in analyzing thermal systems' performance. Several terms encountered in the literature are synonymous or closely related to exergy, which are available energy, essergy, utilizable energy, and availability. The thermal efficiency of the power plants can be increased by adopting supercritical and ultra-supercritical conditions. Pressures exceeding 221.2 bar and steam temperatures exceeding 374.5°C are being used for supercritical thermal power plants, whereas the corresponding values for ultra-supercritical conditions are greater than or equal to 250 bar and 600°C. So, there is ample scope to enhance the steam conditions and optimize power plant configurations further to improve the plant efficiency. The power plant cycle is simulated by using Cycle Tempo version 5.0 software. It is used for advanced thermal energy system optimization. In the present work, for the thermodynamic analysis of supercritical—660 MW, and ultra-supercritical—800 MW power plant, cycle capacities are carried out and compared with subcritical power plant capacity of 500 MW. The important conclusions drawn from the current research work are the improvement in feed water temperature, optimum steam extraction pressures, and comparison of exergy efficiency of various thermal systems.

Keywords: exergy, subcritical (sub), supercritical (SupC), ultra supercritical (USc), exergy loss and exergy efficiency

1. Introduction

Exergy was introduced by Z. Rant in 1956 [1]. It is preferred over energy analysis, because it cannot detect most thermodynamic imperfections, mainly irreversible heat transfer in system components. Exergy offers engineers a refined perspective on energy utilization. In the grand tapestry of engineering principles, the concept of exergy is grounded in understanding that not all energy is created equal [2]. Energy can exist in various forms—thermal, mechanical, chemical—but exergy takes a discerning approach, focusing on the energy that possesses the potential to perform work [3, 4].

2. Energy vs exergy

Energy is a broad measure of the capacity to do work, while exergy narrows the focus to the portion of energy that can be converted into useful work. In simpler terms, energy is like a raw resource, and exergy is the refined, usable portion extracted for practical applications [5].

Exergy is a measure of the quality or usefulness of energy. It is only conserved when there are no irreversible processes in a system and its surroundings, which is never the case [6]. Exergy is always destroyed when irreversibility happens. By doing an exergy analysis on a plant like a power station, a chemical plant, nuclear or a cooling system, we can quantify the thermodynamic inefficiencies as exergy destructions and/or exergy losses, which indicate how much exergy is consumed (or depleted) in the system [7–9]. Exergy analysis considers the different thermodynamic values of different types of energy and amounts, such as work and heat. The exergy transfer with shaft work is the same as the shaft work. The exergy transfer with heat transfer, however, depends on how the temperature of the heat transfer compares to the temperature of the environment [10].

3. The need for exergy analysis

Traditional energy analyses, while informative, may fall short in pinpointing inefficiencies within a system. Exergy analysis, on the other hand, provides a more granular understanding of energy transformations, revealing the areas where exergy losses occur. Considering a steam power plant. Traditional energy analysis would highlight the overall energy input and output, offering insights into the plant's efficiency. However, exergy analysis dives deeper, scrutinizing each component of the system. It identifies inefficiencies in the steam generator, turbine, and condenser, shedding light on specific areas where exergy losses occur [11]. Armed with this knowledge, engineers can implement targeted improvements, such as better insulation or more efficient turbines, to enhance the overall system efficiency. Some of the applications of exergy include:

1. Exergy analysis: This is a method of thermodynamic analysis that uses exergy as a basis for evaluating the performance and improvement potential of energy systems. Exergy analysis can identify the sources and causes of exergy destruction and loss and provide insights for enhancing the system design and operation [12–16].

2. Exergy optimization: This is a method of optimizing the design and operation of energy systems by minimizing the exergy destruction and loss or maximizing the exergy output or efficiency [17, 18]. Exergy optimization can be performed using mathematical models, numerical methods, or heuristic algorithms.

3. Exergy economics: This is a branch of thermoeconomics [19, 20] that uses exergy as a common currency for evaluating the cost and benefit of energy systems. Exergy economics can account for the quality and availability of energy resources, as well as the environmental impact of energy use.

4. Exergy sustainability: This is a concept that relates exergy to the sustainability of energy systems and human activities. Exergy sustainability can be defined as the ratio of exergy input to exergy output, or the ratio of exergy consumption to exergy production. Exergy sustainability can be used to measure the degree of compatibility between energy use and environmental preservation [21–24].

The thermal power is the largest source of India. Coal and lignite [25] have accounted as 60% of India's total installed capacity. India's electricity sector consumption about 72% of the coal produced. In India, the total installation capacity of thermal power production has 207.77 GW. The coal-based power plants are based on the Rankine cycle. The main components of a cycle consist of steam generator, steam turbines, condenser, feed water heaters and several pumps. Power plant optimization can be done by any of the methods below.

1. Steam generator Optimization.
2. Condenser Pressure Reduction.
3. Reduction of steam generator pressure losses and leakages.
4. Minimization of combustion air excess.
5. Thermal losses minimization.
6. Improvement of steam generator efficiency & steam turbine components.
7. Main Steam Parameters Optimization.
8. Proper reheat pressure Optimization.
9. Feed water heater pressure optimization.
10. Steam turbine blade optimization.

3.1 Classification of power plant cycles

The power plants are classified into subcritical, supercritical, and ultra-supercritical, based on operating conditions and plant capacity. The classification of the coal fired power plants, and their steam parameters are summarized [26] shown in **Table 1**. The temperature-entropy plot presentation for subcritical, supercritical, and ultra supercritical cycle is mentioned in **Figure 1**.

3.2 Efficiency of the power plants

Thermal efficiency is very important to economic gain and it's an important engineering effort over 250 years by applying various techniques and technologies implemented in thermal sector [27]. American Electric Power's (AEP), Philo Plant Unit 6, steam generator is the first commercial supercritical unit, in service early in 1957. Philo 6th unit, a double reheat design, delivered 120 MW, operating at a steam flow rate of 85 kg/s, main steam pressure (MSP) 310 bar, MST/SRH/DRH 621/565°C and 538°C. The technology was supplied by the Babcock & Wilcox Company (B&W). In 1959, Philadelphia Electric Company's Eddy stone, steam generator with a double reheat supplied by Combustion Engineering, Inc., delivered the net power output of 325 MW and the net plant thermal efficiency has 39.99% based on the higher heating value (HHV), at a flow rate of 252 kg/s under the main steam pressure of 345 bar, MST/SRH/DRH of 649/565°C and 565°C [28–32]. To investigate the efficiency of power plants by varying the steam parameters, authors have used various modern tools for simulation of

Power plant cycle	Unit size (MW)	Main steam pressure (bar)	Main steam temperature (°C)	Reheat steam temperature (°C)
Subcritical	500	166	538	538
Supercritical	660	247	538	565
Ultra-supercritical	800	270	565	593

Table 1.
Classification of the coal fired power plants.

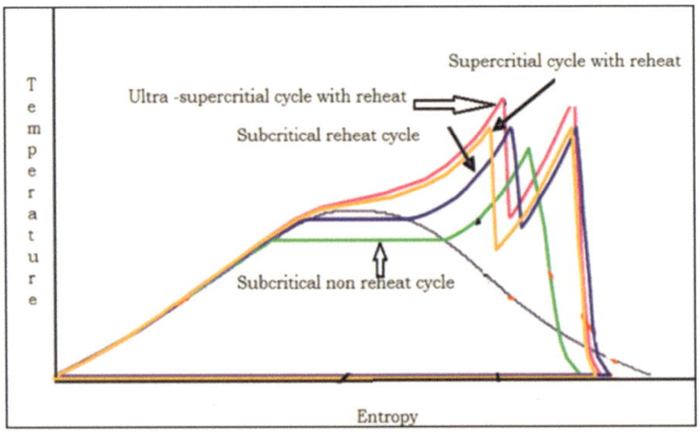

Figure 1.
Temperature – Entropy plane of three different power plant cycles [26].

power plant configurations are discussed in literature reviews are 1. Ebsilon, 2. Cycle Tempo, 3. IAPWS- Microsoft Excel 4. Gate Cycle, 5. Steam Pro, 6. Reference Prop 8.0 7. Engineering Equation Solver (EES), 8. MATLAB, 9. IPSE-PRO.

4. Mathematical modeling equations

Mathematical formulations, pertaining to thermal power plant cycles are presented in terms of exergy analysis. The formulation of power plant processes is inherently non-convex, nonlinear by nature, which means that a good initial starting point and feasible bounds, for variables, must be provided to guarantee a good solution. To explore the benefits of steam parameters and structural changes in the process, it may be possible to build a general layout, containing all the possible process options and subjecting them to optimization. Therefore, it is essential to find a systematic way of building a general layout, which includes all the promising alternatives without being too large. The traditional way of designing power plants is to maximize the thermal efficiency of the plant [33].

4.1 Exergy analysis

Exergy is a property and is associated with the state of the system and the environment. It is not conserved. Once it is wasted, it can never be recovered. The useful work potential of a system is the amount of energy extracted as useful work. The useful work potential of a system at a specified state is called exergy. Exergy destruction or loss is proportional to the increase in entropy of the system together with its surroundings. The loss of exergy or irreversibility provides a quantitative measure of process inefficiency. Exergy efficiency does provide a measure of approach to ideality and exergy loss does provide measures of the deviation from ideality.

Exergy analysis [34–39] is a powerful tool for evaluating the performance of energy conversion processes. It is based on the concept of exergy, which is the maximum amount of work that can be obtained from a system or a flow of matter and energy when it is brought into equilibrium with a reference environment. Exergy analysis allows us to identify the sources and causes of irreversibilities, losses, and inefficiencies in any process, and to quantify them in terms of exergy destruction. Exergy analysis can also help us to design more efficient and sustainable systems, and to optimize the use of available resources.

The exergy rate entering the system is equal to the sum of the exergy rate leaving the system plus the amount of exergy destruction rate during the process.

$$\Sigma Ex_{in} = \Sigma Ex_{out} + \Sigma Ex_{destroyed} \tag{1}$$

The calculation of exergy for different primary energy sources depends on the definition of the reference environment and the thermodynamic properties of the substances involved. For fossil fuels, such as coal, oil, and natural gas, the exergy is equal to their lower heating value, which is the amount of heat released when they are completely burned.

Exergy efficiency is a measure of how effectively a process converts exergy input into exergy output. It is defined as the ratio of exergy output to exergy input, multiplied by 100. Exergy efficiency can be used to compare different processes or systems that have different types and qualities of energy inputs and outputs. Exergy efficiency can also be used to assess the improvement potential of a process or system, by comparing it with an ideal or reversible process that has the same input and output conditions. Exergy efficiency is an important indicator of thermodynamic performance, environmental impact, and economic viability in various fields of industry. The different components of a power plant cycle and its exergy efficiency terms are mentioned below with mathematical equations.

4.1.1 Steam generator

Steam generator is a closed vessel converts water into steam based on the pressure and temperature maintained in the generator by absorbing heat from the burnt coal [38]. The schematic diagram of steam generator shown in **Figure 2**.

Exergy flow of streams for a steam generator can be calculated as.

$$\dot{\psi} = \dot{m}\ x[(h\text{-}h_0)\text{-}T_0\ x(s\text{-}s_0)] \tag{2}$$

Where, h and s represent the specific enthalpy and entropy of steam respectively, 0: dead state condition (atmospheric pressure and temperature).

In the exergetic performance analysis, exergy efficiency gives a measure of the performance of a system or a component.

Exergy absorbed in steam generator.

$$\dot{\psi}_{SG} = \frac{\dot{m}\ x[(h_o\text{-}h_i)\text{-}T_0\ x(s_o\text{-}s_i)]}{\eta_{SG}} \tag{3}$$

Exergy destruction or loss in steam generator can be calculated by

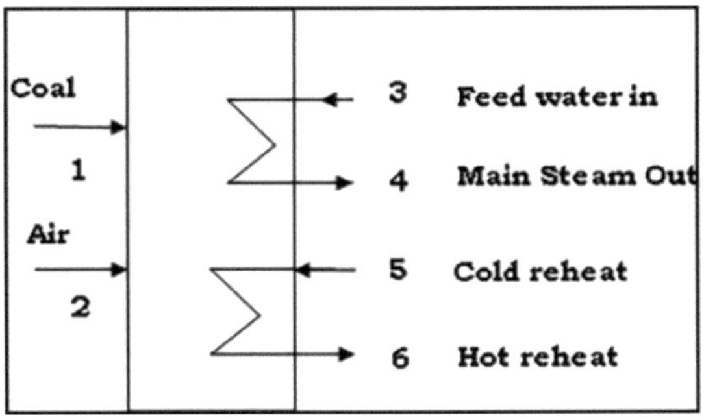

Figure 2.
Schematic diagram of steam generator.

$$\dot{\psi}_{destr,SG} = \dot{\psi}_f - \dot{\psi}_p, \dot{\psi}_f = \dot{\psi}_1 + \dot{\psi}_2 + \dot{\psi}_3 + \dot{\psi}_5, \dot{\psi}_p = \dot{\psi}_4 - \dot{\psi}_6 \quad (4)$$

Where $\dot{\psi}_{destr,SG}$ is the exergy destruction or loss in steam generator, $\dot{\psi}_f$ is called exergy of fuel and it is always destroyed or consumed, whereas $\dot{\psi}_p$ is called exergy of product and it is the amount of exergy which are produced.

$$\text{Exergy efficiency } \varphi_{SG} = ((\dot{\psi}_4 - \dot{\psi}_3) + (\dot{\psi}_6 - \dot{\psi}_5))/(\dot{\psi}_1 + \dot{\psi}_2) \quad (5)$$

Where φ_{SG}: exergy efficiency in steam generator, $\dot{\psi}_1$: Exergy of coal, $\dot{\psi}_2$: Exergy of air, $\dot{\psi}_3$: feed water in, $\dot{\psi}_5$: Cold reheat or steam entering in to the reheater, $\dot{\psi}_4$: Main steam, $\dot{\psi}_6$: Hot reheat or superheated steam outlet from reheater.

4.1.2·Steam turbine

Steam turbine is driven with high pressure steam produced by a steam generator. It is a heat engine which converts the thermal energy of steam into mechanical work [40]. It is used to generate electricity by harnessing the power of steam in a power plant. The schematic diagram of steam turbine is shown in **Figure 3**.

Exergy destruction or loss in steam turbines shown in **Figure 3** can be calculated by.

$$\dot{\psi}_{destr,T} = \dot{\psi}_1 - \dot{\psi}_2 - \dot{\psi}_3 - \dot{W} \quad (6)$$

Where $\dot{\psi}_1$: Exergy of main steam flow in to steam turbine,
$\dot{\psi}_2$, $\dot{\psi}_3$: Steam extractions for feed water heating,
$\dot{W}$: Work output from steam turbine.
Exergy efficiency of the steam turbine can be calculated by.

$$\varphi_T = \frac{\dot{W}}{(\dot{\psi}_1 - \dot{\psi}_2 - \dot{\psi}_3)} \quad (7)$$

4.1.3 Steam condenser

A steam condenser is a heat exchanger applied to convert low pressure exhaust steam from the steam turbine to water. The schematic diagram of steam condenser is shown in **Figure 4**.

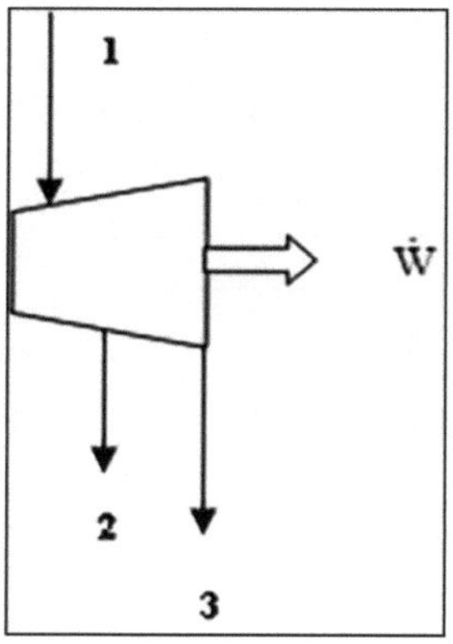

Figure 3.
Schematic diagram of steam turbine.

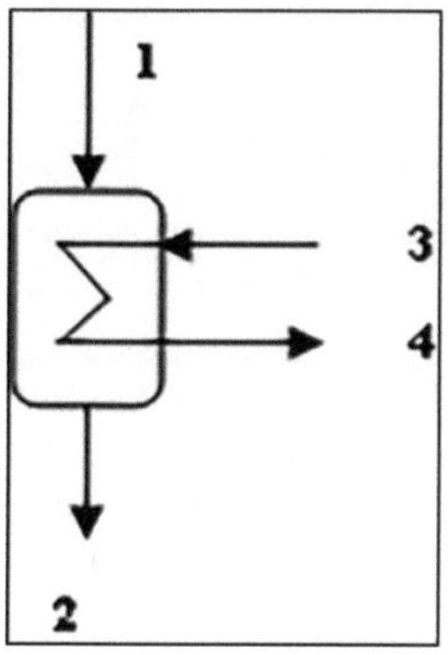

Figure 4.
Schematic diagram of steam condenser.

Exergy destruction or loss in steam condenser shown in **Figure 4** can be calculated as.

$$\dot{\psi}_{destr,COND} = \dot{\psi}_1 + \dot{\psi}_3 - \dot{\psi}_2 - \dot{\psi}_4 \tag{8}$$

Exergy efficiency in steam condenser can be calculated by Eq. (9)

$$\varphi_{COND} = \frac{(\dot{\psi}_4 - \dot{\psi}_3)}{(\dot{\psi}_1 - \dot{\psi}_2)} \tag{9}$$

4.1.4 Feed water heater

A feedwater heater is a power plant component used to preheat water delivered to a steam generator. This process helps to improve the steam generator efficiency and reduces the fuel costs. The schematic diagram of feedwater is shown in **Figure 5**.

Exergy destruction or loss in feed water heater shown in **Figure 5** calculated as.

$$\dot{\psi}_{destr,FWH} = \dot{\psi}_1 + \dot{\psi}_3 - \dot{\psi}_2 - \dot{\psi}_4 \tag{10}$$

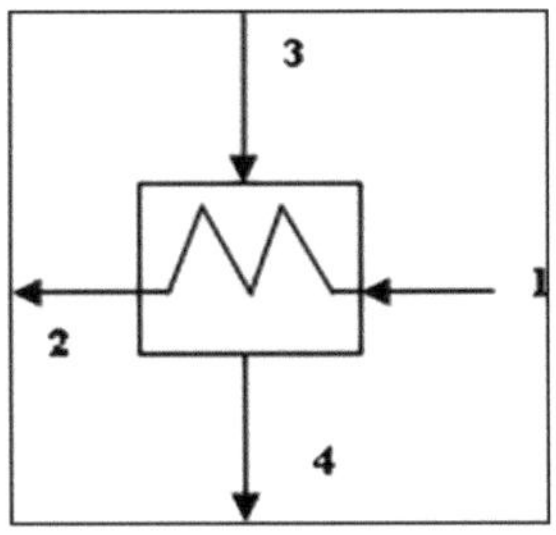

Figure 5.
Schematic diagram of one feed water heater.

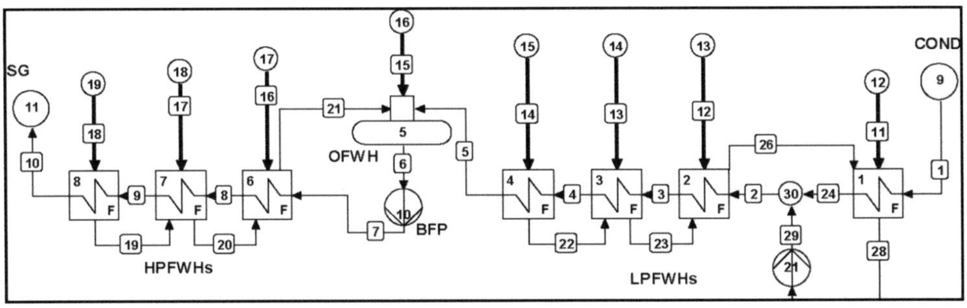

Figure 6.
Schematic layout of eight feed water heaters (LPFWHs and HPFWHs).

$$\dot{m}_1 + \sum_{x=11}^{18} \dot{m}_x = \dot{m}_{10} + \dot{m}_{26} \tag{11}$$

$$\dot{m}_1 h_1 + \sum_{x=11}^{18} \dot{m}_x h_x - \dot{m}_{10} h_{10} - \dot{m}_{26} h_{26} = \dot{Q}_{FWH} \tag{12}$$

$$\dot{m}_1 \psi_1 + \sum_{x=11}^{18} \dot{m}_x \psi_x - \dot{m}_{10} \psi_{10} - \dot{m}_{26} \psi_{26} - \dot{\psi}_{\dot{Q}_{FWH}} = \dot{\psi}_{dest,FWH} \tag{13}$$

A power plant cycle consists of open feed water heaters (OFWH) called direct contact heat exchanger and closed feed water heaters called indirect contact heat exchanger. The closed feed water heaters are classified as low pressure closed feed water heaters (LPFWHs) and high-pressure feed water heaters (HPFWHs). The layout of different feed water heaters is shown in **Figure 6**. Part of steam is extracted from various steam turbines called low pressure, intermediate and high-pressure steam turbine to raise the condensed water temperature from steam condenser.

Exergy efficiency for high pressure feed water heaters φ_{FWH} calculated as

$$\frac{\dot{\psi}_{10} - \dot{\psi}_9}{\dot{\psi}_{18} - \dot{\psi}_{19}} \tag{14}$$

$$\text{HPFWH2} = \frac{\dot{\psi}_9 - \dot{\psi}_8}{\dot{\psi}_{17} + \dot{\psi}_{19} - \dot{\psi}_{20}} \tag{15}$$

$$\text{HPFWH1} = \frac{\dot{\psi}_8 - \dot{\psi}_7}{\dot{\psi}_{16} + \dot{\psi}_{20} - \dot{\psi}_{21}} \tag{16}$$

$$\text{For Open feed water heaters (DEA)} = \frac{\dot{\psi}_6}{\dot{\psi}_{15} + \dot{\psi}_5 + \dot{\psi}_{21}} \tag{17}$$

Exergy efficiency φ_{FWH} for low pressure feed water heaters calculated as

$$\text{LPFWH 4} = \frac{\dot{\psi}_5 - \dot{\psi}_4}{\dot{\psi}_{14} - \dot{\psi}_{22}} \tag{18}$$

$$\text{LPFWH 3} = \frac{\dot{\psi}_4 - \dot{\psi}_3}{\dot{\psi}_{13} + \dot{\psi}_{22} - \dot{\psi}_{23}} \tag{19}$$

$$\text{LPFWH 2} = \frac{\dot{\psi}_3 - \dot{\psi}_2}{\dot{\psi}_{12} + \dot{\psi}_{23} - \dot{\psi}_{26}} \tag{20}$$

$$\text{LPFWH 1} = \frac{\dot{\psi}_{24} - \dot{\psi}_1}{\dot{\psi}_{11} + \dot{\psi}_{26} - \dot{\psi}_{28}} \tag{21}$$

4.1.5 Feed water pump

Feed water pump is a power plant component used to pump the feed water coming from various feed water heaters into the steam generator and its schematic diagram is shown in **Figure 7**.

$$\text{Exergy destruction or loss in pump } \dot{\psi}_{destr,P} = \dot{\psi}_1 \text{-} \dot{\psi}_2 + \dot{W}_p \tag{22}$$

$$\text{Exergy efficiency in pump } \varphi_P = \frac{(\dot{\psi}_2 \text{-} \dot{\psi}_1)}{\dot{W}_p} \tag{23}$$

Total exergy destruction or loss in the plant can be determined as sum of exergy destruction rates of components:

$$\dot{\psi}_{destr,total} = \sum \dot{\psi}_{destr,i} = \dot{\psi}_{destr,SG} + \dot{\psi}_{destr,T} + \dot{\psi}_{destr,C} + \dot{\psi}_{destr,P} + \dot{\psi}_{destr,FWH} \tag{24}$$

For the whole thermal power plant, the exergy efficiency can be given.

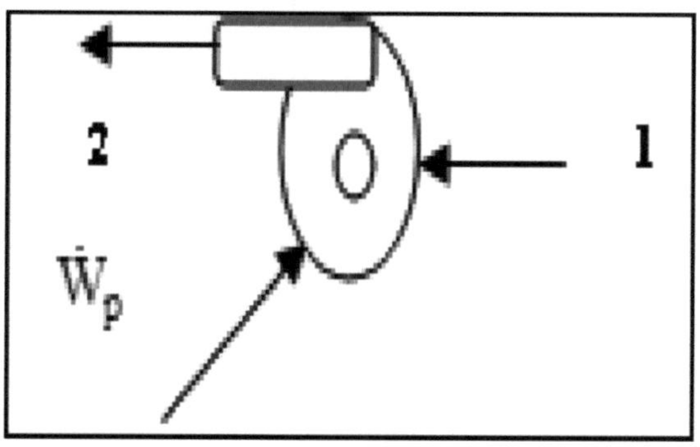

Figure 7.
Schematic diagram of feed water pump.

$$\varphi = \frac{\dot{W}_{net}}{\dot{m}_{coal} x \psi_{coal}} \quad (25)$$

Where ψ_{coal}: specific exergy of coal, $\dot{m}_{coal}$: Mass of the coal.
$\dot{W}_{net}$: Network output from turbines.

4.2 Quasi-Newton optimization

The process to get maximum thermal efficiency of a cycle describes the optimization steps in Cycle Tempo. Quasi-Newton optimization technique in Cycle Tempo 5.0 is recognized today as one of the most powerful tools for feeding steam extraction pressures optimization from steam turbines. It usually converges fast, and sometimes converges even without step length control. In the initial step, give the input values to the components presented in the cycle diagram. In the second step, start with initial guess values by giving starting steam extraction pressures from turbines X_i, Let i = 1. In the third step, it calculates the mass & energy balance equations. In the fourth step, it computes the objective function $f(X_i)$ as η of thermal power plant cycle. In the fifth step, it generates a new solution X_{i+1} as steam extraction pressures from the turbines. In the sixth step, it computes the objective function $f(X_{i+1})$ as thermal efficiency (η) of power plant. In the seventh step, it checks for convergence on efficiency of plant. In the eighth step, it will check for maximum efficiency of a plant. In the ninth step, if it is ok that gives the message as success and updates the geometry as an optimal solution.

5. Result and discussions – Case study

The simulation and optimization of three different thermal power plant cycles are carried out, with Cycle Tempo 5.0 simulation software tool for energy systems. The following parameters have been studied for 500, 660 and 800 MW power plant cycles. The performance of the power plant cycle configuration is mentioned in **Tables 2-6** and shown in **Figures 8-12**. In the current study exergy analysis was used as a tool to identify the thermodynamic losses and exergy efficiencies in various components of the power plant cycle with the help of Cycle Tempo 5.0. [41].

The key output parameters which can be observed to see for changed input parameters are exergy absorbed in steam generator and reheater, Exergy loss/Total exergy input in steam generator, reheater and steam turbines, Exergy loss in different components, Exergy loss and rate of heat transfer in closed regenerative feed water heaters, Thermal and exergy efficiency, Impact of condenser pressure variation on exergy efficiency, Auxiliary power consumption and pipe exergy losses, Coal, and steam consumption. All these parameters are based on steam extraction pressure optimization from feed water heaters to get maximum plant cycle efficiency.

Efficiency of the Rankine cycle can be improved by increasing the average temperature of heat addition or by decreasing the average temperature of heat rejection [42].

In thermal power plants, the average heat addition is increased, by setting number of feed water heaters. The feed water enters the steam generator (Evaporator), through economizer. The performance of a power plant is characterized by its power

MSP (bar)	Net thermal and exergy efficiency (%)					
	500°C		600°C		700°C	
200	41.33	38.96	42.30	39.88	43.59	41.09
220	41.49	39.11	42.85	40.39	44.09	41.56
240	41.60	39.21	43.03	40.56	44.29	41.75
260	41.67	39.28	43.16	40.69	44.47	41.92
280	41.71	39.32	43.27	40.79	44.62	42.06
300	41.70	39.31	43.33	40.86	44.79	42.13

Table 2.
Net thermal and exergy efficiency of plant (%) at different main steam temperatures and pressures.

Main steam pressure (bar)	Main steam temperature		
	500°C	600°C	700°C
200	752.54	677.73	603.51
250	810.35	696.86	614.06
300	847.91	718.93	631.77

Table 3.
Variation of steam consumption (kg/s) at different main steam pressures and temperatures.

Main steam pressure (bar)	Main steam temperature		
	500°C	600°C	700°C
200	453	443	430
250	452	438	422
300	450	433	418

Table 4.
Variation of coal consumption in tonnes per hour at different main steam pressures and temperatures.

MST	Subcritical	Supercritical	Ultra-supercritical
	Auxiliary power consumption (MW)		
500°C	6.65	8.77	10.64
600°C	6.17	8.12	9.54
700°C	5.75	7.84	8.75

Table 5.
Comparative auxiliary power consumption in three different power plant cycles.

Feed water heaters	Subcritical	Supercritical	Ultra-supercritical
	Steam extraction pressures (bar)		
LPFHW1	0.32	0.34	0.36
LPFWH2	0.87	0.96	0.98

Feed water heaters	Subcritical	Supercritical	Ultra-supercritical
	Steam extraction pressures (bar)		
LPFWH3	1.96	2.26	2.50
LPFWH4	4.04	4.78	5.78
DEA(OFWH)	7.96	9.59	12.01
HPFWH1	14.62	18.35	23.34
HPFWH2	31.21	39.80	48.94
HPFWH3	75.61	92.31	107.58
Feed Water temperature (°C)	290.65	304.40	315.34

Table 6.
Comparison of steam extraction pressure (bar) optimization in three different power plant cycles.

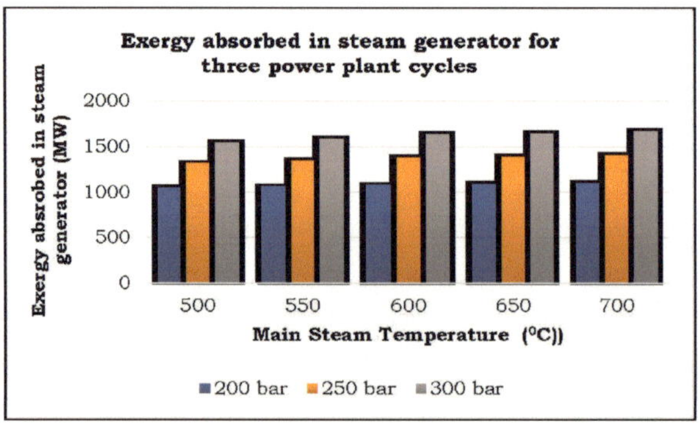

Figure 8.
Exergy absorbed in steam generator for three power plant cycles.

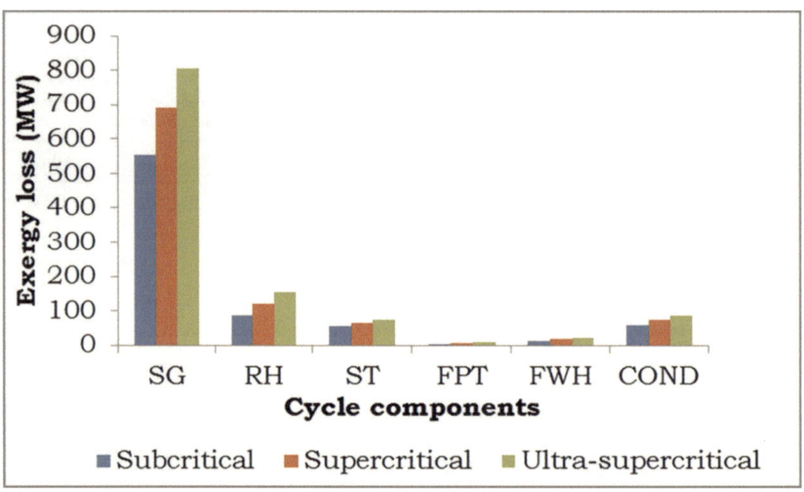

Figure 9.
Exergy loss of different components in three different power plants.

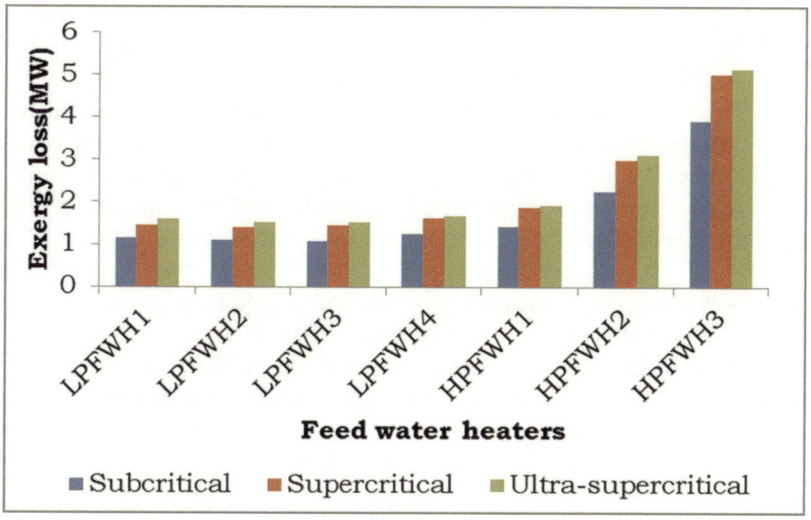

Figure 10.
Exergy loss in closed feed water heaters of three different power plant cycles.

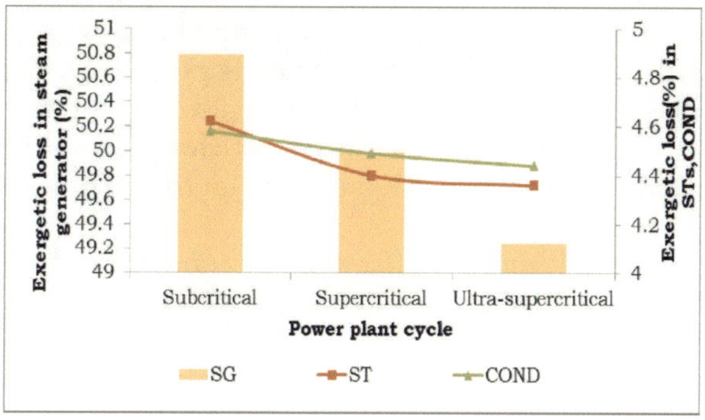

Figure 11.
Exergetic loss variations of different components of three different power plant cycles.

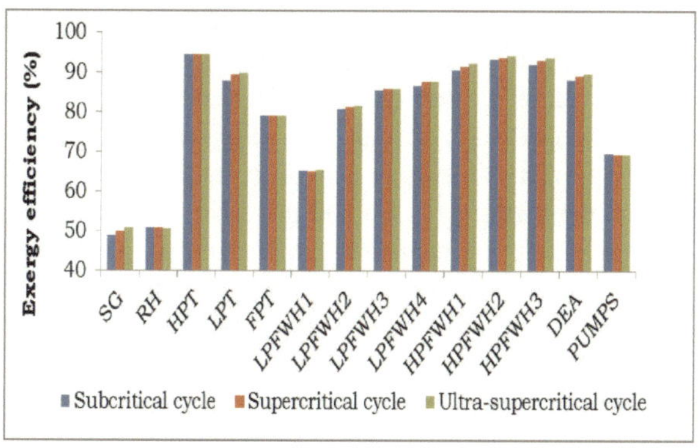

Figure 12.
Variation of exergy efficiency of power plant components in three different power plant cycles.

output, efficiency, heat rate, steam rate, fuel rate, etc., In the present work, the efficiency of thermal power plant cycles (subcritical, supercritical and ultra-supercritical) is analyzed by considering, the key input parameters of, main steam pressure (MSP), main steam temperature (MST), reheat temperature and condenser pressure, etc., **Figure 13** shows, the schematic layout of thermal power plant, contains combustor (air + coal) releases the flue gases and passes through the various heat exchangers like (evaporator, superheaters), where feed water gets converted to steam according to the steam parameters maintained in a power plant cycle called as steam generator (SG). The steam and then passes into the high-pressure turbine (HPT), reheater to low pressure turbine (LPT), condenser (COND), four closed feed water heaters (four low pressure feed water heaters (LPFWHs) and three high pressure feed water heaters (HPFWHs)), one open feed water heater (OFWH) called as deaerator (DEA) called feed water heater section and various pumps (Boiler feed pump, water pumps). The feed water from feed water heater section into the economizer and then passes into the evaporator, where it gets converted into steam based on steam parameters.

The assumptions considered for the thermal power plant cycle shown in **Figure 13** are available in **Table 7**. Before the analysis, any power plant components should be given some assumptions to evaluate the performance of the system.

Table 2 shows the effect of main steam pressure on thermal and exergy efficiency of a plant cycle. It shows that the exergy efficiency of a cycle is less than thermal efficiency of cycle and this is due to chemical exergy of coal is greater than the calorific value of coal. As the main steam pressure is raised, the thermal efficiency increases up to a pressure of 280 bar for subcritical power plant cycle and there after it decreases. This shows that the optimum limit of efficiency increases. After 280 bar, the efficiency decreases due to less work output from the turbines and increase of pump work.

Variation of steam consumption (kg/s) at different main steam pressures and temperatures are shown in **Table 3**. It shows the steam consumption increases when the main steam pressure is raised due to reduced enthalpy of steam. Steam consumption decreases when the main steam temperature is raised at a particular pressure of 200 bar due to high enthalpy of steam. To meet the required power plant capacity of 500MW with high enthalpy steam content can reduce steam consumption.

The coal consumption in tonnes per hour (TPH) for a power plant operating at different steam conditions is shown in **Table 4**. It shows that coal consumption decreases, both with the increase of main steam pressures and temperatures. But the decrease of coal consumption is little bit less when main steam pressure is raised as compared to the different main steam temperatures.

The comparison of auxiliary power consumption results is shown in **Table 5**. The auxiliary power consumption decreases, as the main steam temperatures have increased from 500–700°C. The auxiliary power consumption decreases to all the power plant cycles, because it depends on the specific volume of steam and quality of steam entering pumps when the main steam temperature is raised. The cooling water required by the condenser also decreases when main steam temperatures are raised. Auxiliary power consumption increases, with the increase of the power plant capacity.

The percentage of auxiliary power consumption has increased 24%, compared to subcritical plant, whereas the same value is increased to 14.8%, compared to supercritical power plant cycle. But overall, 35% auxiliary power consumption is increased compared to subcritical power plant cycle.

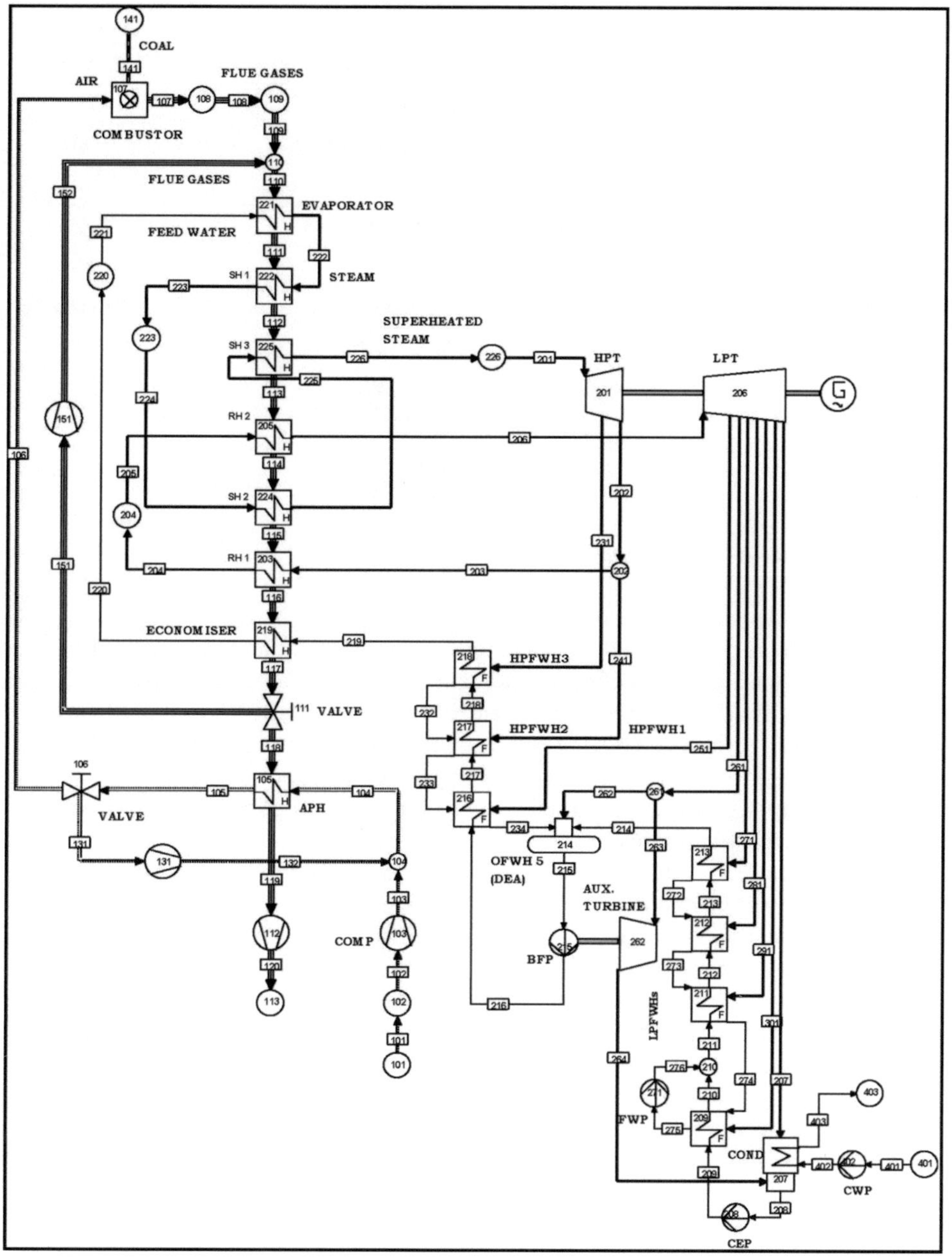

Figure 13.
Process flow sheet layout of thermal power plant cycle [41].

S. no	Component	Value
1	Pressure drops in steam generator	3%
2	Efficiency of steam generator	92%
3	Pressure drops in reheater	1%
4	Pressure drops of main steam	4%

S. no	Component	Value
5	Isentropic efficiency of high-pressure turbine	90%
6	Isentropic efficiency of a low-pressure turbine	90%
7	Mechanical efficiency of a feed pump turbine	98%
8	Isentropic efficiency of a feed pump	78%
9	Isentropic efficiency of a main condensate pump	78/75%
10	Generator efficiency	98%
11	Atmospheric temperature/pressure	30°C/1.01 bar
12	TTD on high side/low side for LPFWHs	3/5 K
13	TTD on high side/low side for HPFWHs	0/7 K
14	Condenser pressure	0.09 bar
15	Calorific value of coal	34,000 kJ/kg

Table 7.
Assumptions of a power plant components [41].

The comparison of steam extraction pressure optimization of 8 regenerative feed water heater stages and feed water temperatures, in the power plant cycles are shown in **Table 6**. It shows that the optimum steam extraction pressures and feed water temperatures are increased. This is due to the increase in steam flow rate. The feed water temperature improvement is 4.51%, compared to subcritical steam power plant cycle, 3.47% compared to supercritical power plant cycle and overall, 7.82% compared to subcritical power plant cycle.

The variation of exergy absorbed in steam generator is shown in **Figure 8**. It shows the exergy absorbed in steam generator increases gradually, as the main steam temperature is raised. This is due to the increase in entropy of steam.

Exergy is how much work output is possible to convert on using heat content of steam. But overall, the exergy absorbed both in steam generator and reheater decreases due to constant reheat temperature. The exergy absorbed in steam generator is 21.47% compared to subcritical power plant, 15.46% compared to supercritical, and overall, the exergy absorbed in steam generator is 33.62% between ultra-supercritical to subcritical cycle at a main steam temperature of 600°C.

The exergy loss, measured in megawatt (MW) of different plant cycle components, for three different power plants capacity are shown in **Figure 9**. It shows the exergy loss increases to all components. The major exergy loss found in steam generator but in reheater, steam turbine and in condenser it decreases. When the power plant capacity rises the difference in exergy loss increases. The difference in exergy loss in terms of percentage increases in steam generator between subcritical and supercritical up to 19.87%. Whereas the same value in between the supercritical and ultra-supercritical is 41.10%. The overall difference in exergy loss is 31.17% between the subcritical and ultra-supercritical. The difference in exergy loss in percentage is increased in reheater between subcritical and supercritical is 27.10% whereas the same value in between the supercritical and ultra-supercritical is 21.67%. The overall difference in percentage exergy loss between the subcritical and ultra-supercritical is 42.90%.

The difference in exergy loss in percentage is increased in feed pump turbine between subcritical and supercritical is 40.70% whereas the same value in between the supercritical and ultra-supercritical is 33.36%. The overall difference in percentage exergy loss between the subcritical and ultra-supercritical is 60.48%. The difference in exergy loss in terms of percentage found less in feed water heater and then in condenser.

The exergy loss in feed water heaters is up to 8 regenerative stages which are shown in **Figure 10**. As the number of feed water heaters increases the exergy loss decreases. The difference in exergy loss percentage is increased in regenerative feed water heaters between subcritical and supercritical is 22.92%. The same value between the supercritical and ultra-supercritical is 4.47%. The overall difference in exergy loss (%) between the sub and ultra-supercritical is 26.37%.

The percentage exergetic loss in steam generator, steam turbines and condenser decrease from subcritical to ultra-supercritical power plant cycles shown in **Figure 11**. The major percentage exergetic loss occurs in steam generator and its value in sub-critical state is 50.79% and the same value in supercritical state is 49.99%. It shows the percentage of exergetic loss decreases in between subcritical to supercritical is 1.60% and the same in between supercritical and ultra-supercritical is only 1.52%.

The overall percentage difference between ultra-supercritical and subcritical is 3.14%. The exergetic loss also decreases in steam turbines and in condenser. But the overall reduction in percentage is very small.

The exergy efficiency of individual components like steam generator, reheater, different turbines, feed water heaters and pumps are shown in **Figure 12**. It shows the exergy efficiency increases in all the components except feed pump. The energy efficiency of steam generator and reheater lie in the range of 48 to 51%. HPT and HPFWHs show the high exergy efficiency in the range of 90 to 95%. LPT and LPFWHs show the exergy efficiency in the range of 80 to 90% and feed pump shows the exergy efficiency in the range of 65 to 70%. Only the feed pump turbine has constant exergy efficiency in all power plant cycles. The power producing device shows more exergy efficiency. The feed water heaters have more exergy efficiency because of the heat transfer between steam and feed water. In deaerators the exergy efficiency decreases as the main steam temperature have raised. But when main steam pressure is raised, the exergy efficiency of deaerator increases.

6. Conclusions

From the aforementioned information in tables and figures, the following conclusions may be drawn as

1. As the main steam temperature is raised by keeping reheat temperature as constant the optimum steam extraction pressure values for 8 regenerative stages decreases.

2. Auxiliary power consumption decreases as the main steam temperature is raised for a main steam pressure to all power plant capacities.

3. The energy and exergy absorbed in steam generator increases to three power plant capacities.

4. Exergy loss in steam generator decreases from subcritical power plant cycle to ultra-supercritical power plant cycle.

5. Comparing the rates of heat transfer in feed water heaters from subcritical to ultra-supercritical power plant cycle, more heat transfer is possible between

subcritical to supercritical compared to supercritical and ultra-supercritical steam conditions.

6. The exergy efficiency of a steam generator increases as the main steam input parameters are raised.

7. The feed water temperature of a plant can be improved by raising the steam input pressures.

8. The exergy analysis reveals that the subcritical units are less efficient to utilize the available exergy of coal compared with supercritical and ultra-supercritical units.

9. Specific steam and coal consumption decreases when higher steam input parameters are used.

10. Simulation and exergy analysis are useful in system optimization. They not only indicate the direction of system optimization, but also provide a theoretical basis for the system optimization.

11. The average cooling water requirement in condenser/MW for subcritical plant capacity of 500 MW is 40 kg/s, for supercritical 660 MW, it is 38 kg/s and for ultra-supercritical 800 MW capacity, it is 37 kg/s.

Acknowledgements

The authors would like to thank the ASIMPTOTE – Cycle Tempo 5.0 licensed software, Delft University of technology, Netherlands for carrying out the exergy analysis of thermal power plants in the year 2016–2017.

Conflict of interest

The authors declare no conflict of interest.

Nomenclature

$\dot{\psi}$	exergy flow rate (MW)
$\dot{Q}$	heat flow rate (MW)
$\dot{W}$	work (MW)
ψ	specific exergy (kJ/kg)
h	enthalpy (kJ/kg)
LHV	lower heating value (kJ/kg)
HHV	higher heating value (kJ/kg)
$\dot{m}$	mass flow rate (kg/s)
$\dot{m}_s$	mass flow rate of steam (kg/s)
s	entropy (kJ/kg K)

T	temperature (K or °C)
η_{th}	thermal efficiency (%)
φ	exergy efficiency (%)
$\dot{\psi}_f$	exergy of the fuel (MW)
$\dot{\psi}_p$	exergy of the product (MW)

Subscripts

a	air
aux, ele	auxiliary electrical equipment
ext	extraction
destr	destruction or loss
f	fuel
fw	feed water
FWH	feed water heater
gross	gross
loss	loss in components
ms	main steam
net	net
pump	pump
rs	reheated steam
SG	steam generator
COND	condenser
i	inlet
o	outlet
p	product
rh	reheat
S	steam
T	turbine
total	total
th	thermal

Abbreviations

APC	auxiliary power consumption
BFP	boiler feed pump
BFPT	boiler feed pump turbine
CEP	condensate extraction pump
COND	condenser
CT	cooling tower
CWP	cooling water pump
DEA	deaerator
FPT	feed pump turbine
FWH	feed water heater
FWT	feed water temperature
GW	Gigawatt
HHV	higher heating value
HPFWH	high pressure feed water heaters
HPT	high pressure turbine

LHV	lower heating value
LPFWH	low pressure feed water heater
LPT	low pressure turbine
MW	Megawatt
MSP	main steam pressure
MST	main steam temperature
OFWH	open feed water heater (DEA)
SG	steam generator
SRH	single reheater
TBFP	turbine driven boiler feed pump
TPH	tonnes per hour
TTD	terminal temperature difference

Author details

Pasupuleti Ravindra Kumar[1*] and Naradasu Ravi Kumar[2]

1 Lakireddy Bali Reddy College of Engineering, Mylavaram, Andhra Pradesh, India

2 MVGR College of Engineering, Vizianagaram, Andhra Pradesh, India

*Address all correspondence to: pasupuletirk@gmail.com

References

[1] Moran MJ, Shapiro HN. Fundamentals of Engineering Thermodynamics. 3rd ed. New York: John Wiley & Sons; 1996

[2] Cengel YA, Boles MA. Thermodynamics: An Engineering Approach. 5th ed. Boston: McGraw-Hill; 2006

[3] Khartchenko NV, Kharchenko VM. Advanced Energy Systems. 2nd ed. Boca Raton, London, Newyork: CRC Press, Taylor & Francis group; 2013. p. 97. Availbale from: https://books.google.com/books/about/Advanced_Energy_Systems_Second_Edition.html?id=BSAtAgAAQBAJ

[4] Sonntag R, Borgnakke C, Van Wylen GJ. Fundamentals of Thermodynamics. 6th Wiley student ed. University of Michigan: John Wiley and Sons Inc.; 2003. pp. 302-363. Availbale from: https://bcs.wiley.com/he-bcs/Books?action=index& bcsId=1252& item Id=0471152323. ISBN: 0-471-152 32-3

[5] Kanoglu M, Dincer I, Rosen MA. Understanding energy and exergy efficiencies for improved energy management in power plants. Energy Policy. 2007;**35**:3967-3978. DOI: 10.1016/j.enpol.2007.01.015

[6] Bejan A. Fundamentals of exergy analysis, entropy generation minimization, and the generation of flow architecture. International Journal of Energy Research. 2002;**26**(7):545-565. DOI: 10.1002/er.804

[7] Rosen MA. Clarifying thermodynamic efficiencies and losses via exergy. International Journal of Exergy. 2002;**2**:3-5. DOI: 10.1016/S1164-0235(02)00055-9

[8] Dincer I. The role of exergy in energy policy making. Energy Policy. 2002; **30**(2):137-149. DOI: 10.1016/S0301-4215(01)00079-9

[9] Rosen MA. Energy- and exergy-based comparison of coal-fired and nuclear steam power plants. Exergy. 2001;**1**(3): 180-192. DOI: 10.1016/S1164-0235(01)00024-3

[10] Rosen M, Dincer I. Effect of varying dead-state properties on energy and exergy analyses of thermal systems. International Journal of Thermal Sciences. 2004;**43**:121-133. DOI: 10.1016/j.ijthermalsci.2003.05.004

[11] Tsatsaronis G, Cziesla F. Basic exergy concepts, exergy balance and exergetic efficiency, exergy analysis of simple processes, energetic and exergetic analysis of complex systems, strength and limitations of exergy analysis. In: Encyclopedia of Life Support Systems (EOLSS), Topic Energy, Developed under the Auspices of the UNESCO. Oxford, UK: EOLSS Publishers; 2004. Available from: http://www.eolss.net

[12] Horlock JH et al. Exergy analysis of modern fossil-fuel power plants. ASME Journal of Engineering for Gas Turbines and Power. 2000;**122**:1-7. DOI: 10.1115/1.483170

[13] Aljundi IH. Energy and exergy analysis of a steam power Plant in Jordan. Applied Thermal Engineering. 2009;**29**(2–3):324-328. DOI: 10.1016/j.applthermaleng.2008.02.029

[14] Gupta MK, Kaushik SC. Exergy analysis and investigation for various feed water heaters of direct steam generation solar – Thermal power plant. Renewable Energy. 2010;**35**:1228-1235. DOI: 10.1016/j.renene.2009.09.007

[15] Amirabedin E, Zeki Yilmazoglu M. Design and exergy analysis of a thermal power plant using different types of Turkish lignite. International Journal of Thermodynamics (IJoT). 2011;**14**(3): 125-133. DOI: 10.5541/ijot.288

[16] Suresh MVJJ, Reddy KS, Kolar AK. 3-E analysis of advanced power plants based on high ash coal. International Journal of Energy Research. 2010;**34**: 716-735. DOI: 10.1177/0957650911418421

[17] Sanpasertparnich T, Aroonwilas A. Simulation and optimization of coal-fired power plants. Energy Procedia. 2009;**1**(1):3851-3858. DOI: 10.1016/j.egypro.2009.02.187

[18] Witold E, Łukasz K, Maciej M. Numerical thermodynamic optimization of supercritical coal fired power plant with support of IPSEpro software. Archives of Thermodynamics. 2012; **33**(3):101-110 Available from: https://bibliotekanauki.pl/articles/240118.pdf

[19] Saidur R, Ahamed JU, Masjuki HH. Energy exergy and economic analysis of industrial boilers. Energy Policy. 2010; **38**:2188-2197. DOI: 10.1016/j.enpol.2009.11.087

[20] Wang L, Yang Y, Changqing D, Yang Z, Xu G, Wu L. Exergo-economic evaluation of a modern ultra-supercritical power plant. Energies. 2012;**5**(6):3381-3397. DOI: 10.3390/en5093381

[21] Kosowski K et al. Steam and Gas Turbines with Examples of Alstom Technology. 2nd ed. Three parts in one volume. Alstom, France, Switzerland, United Kingdom, Poland; 2007

[22] Kuprianov VI. Applications of a cost-based method of excess air optimization for the improvement of thermal efficiency and environmental performance of steam boilers. Renewable and Sustainable Energy Reviews. 2005;**9**:474-498. DOI: 10.1016/j.rser.2004.05.006

[23] Singh OK, Kaushik SC. Variables influencing the exergy-based performance of a steam power plant. International Journal of Green Energy 2013;**10**(3):257-284. DOI: 10.1080/15435075.2011.653847

[24] Siva RV, Kaushik SC, Tyagi SK. Exergetic analysis and evaluation of coal fired supercritical thermal power plant and natural gas fired combined cycle power plant. Clean Technology Environmental Policy. 2013;**15**:1-11. DOI: 10.1007/s10098-013-0647-x

[25] Ganapathy T, Alagumurthi N, Gakkhar RP, Murugesan K. Exergy analysis of operating lignite fired thermal power plant. Journal of Engineering Science and Technology Review. 2009;**2**(1):123-130. DOI: 10.25103/jestr.021.23

[26] Anto B, Hasan MM. Analysis of Supercritical technology in Indian Environment and Utilizing Indian coal. 2010. pp 1-163. Availbale from: https://www.scribd.com/doc/20902219/Paper-on-Super-Critical-Technology-and-Analysis-for-Indian-Environment

[27] Srinivas T, Gupta AVSSKS, Reddy BV. Generalized thermodynamic analysis of steam power cycles with 'n' number of feed water heaters. International Journal of Thermodynamics. 2007;**10**(4):177-185. DOI: 10.5541/ijot.201

[28] Weizhong F. 1000MW ultra-supercritical turbine steam parameter optimization. Frontiers of Energy and Power Engineering in China. 2008;**2**(2):

187-193. DOI: 10.1007/s11708-008-0030-5

[29] Yanjun F, Wang Z. Mathematical modelling and dynamic simulation of intermediate point temperature for ultra-supercritical once-through boiler II. In: IEEE Transaction 2012, The 6th International Conference on Soft Computing and Intelligent Systems, and The 13th International Symposium on Advanced Intelligence Systems. Vol. 2. 2012. pp. 922-927. DOI: 10.1109/SCIS-ISIS.2012.6505267

[30] Zhou L, Gang X, Cheng X, Yongping Y, Li Y, Jianling D. Optimum design of a double reheat steam system in an ultra-supercritical unit. In: Proceedings of ASME Turbo Expo 2013: Turbine Technical Conference and Exposition, June 3-7. Vol. 5(B). San Antonio, Texas, USA; 2013. pp. 1-11. DOI: 10.1115/GT2013-95190. ISBN: 978-0-7918-5520-1

[31] Yongping Y, Ligang W, Changqing D, Xu G, Morosuk T. Comprehensive exergy-based evaluation and parametric study of a coal fired ultra-supercritical power plant. Applied Energy. 2013;**112**:1087-1099. DOI: 10.1016/j.apenergy.2012.12.063

[32] Li Y, Luyao Z, Gang X, Yaxiong F, Shifei Z, Yongping Y. Thermodynamic analysis and optimization of a double reheat system in an ultra-supercritical power plant. Energy. 2014;**74**: 202-214. DOI: 10.1016/j.energy.2014.05.057

[33] Carl KF. Modeling, Analysis and Optimization of Process and Energy systems, Louisiana State University Baton Rouge, LAA: John Wiley & Sons, Inc.; 2012. pp. 397-417. Availbale from: https://onlinelibrary.wiley.com/doi/book/10.1002/9781118121160

[34] Li Z, Li Z, Yan Z. Energy and exergy analysis for three type 500MW steam power plants. Applied Mechanics and Materials. 2012;**148–149**:1131-1136 Available from: www.scientific.net

[35] Erdem H, Akkaya AV, Dagdas A, Sevilgen SH, Sahin B, Tek I, et al. Comparative energetic and exergetic performance analyses for coal-fired thermal power plants in Turkey. International Journal of Thermal Sciences. 2012;**48**(11):2179-2186. DOI: 10.1016/j.ijthermalsci.2009.03.007

[36] Mitrovic D, Zivkovic D, Lakovic MS. Energy and Exergy Analysis of 348.5 MW Steam Power Plant. Vol. 32. Part A. Taylor & Francis; 2010. pp.1016-1027. DOI: 10.1080/15567030903097012

[37] Kelly S, Tsatsaronis H, Morosuk T. Advanced exergetic analysis: Approaches for splitting the exergy destruction into endogenous and exogenous parts. Energy. 2009;**34**(3): 384-391. DOI: 10.1515/tjj-2016-0074

[38] Regulagadda P, Dincer I, Naterer GF. Exergy analysis of a thermal power plant with measured boiler and turbine losses. Applied Thermal Engineering. 2010;**30**:970-976. DOI: 10.1016/j.applthermaleng.2010.01.008

[39] Sandhya H, Aroonwilas A, Veawab A. Exergy analysis of Ultra-supercritical power plant. Energy Procedia. 2013;**37**:2544-2551. DOI: 10.1016/j.egypro.2013.06.137

[40] Leizerovich AS. Steam Turbines for Modern Fossil-fuel Power Plants. 1st ed. New York: River Publishers; 2008. DOI: 10.1201/9781003151388

[41] Asimptote bv.De Schans 23; 5473 PH Heeswijk-Dinther; The Netherlands.

Available from: https://asimptote.com/cycle-tempo/

[42] Madejski P. Introductory Chapter: New Trends and Recent Developments for Thermal Power. InTechOpen; 2018. DOI: 10.5772/intechopen.74723. Available from: https://www.intechopen.com/chapters/59649

Chapter 4

Exergy and Quantum Batteries

Farzaneh Hatami Kamin and Shahriar Salimi

Abstract

The study of quantum thermodynamics has led to the development of quantum batteries. These devices use quantum advantages to store and extract useful energy from physical systems. Ergotropy is the maximum work that can be extracted from a quantum system by cyclic unitary operations. When external thermal baths couple with the quantum battery, there is energy loss due to thermal effects on the system. In some cases, a part of the total energy available in the system cannot be stored as ergotropy. Therefore, it is important to consider the amount of residual energy that cannot be extracted as useful work from quantum batteries by unitary processes. To better understand the amount of energy lost during work extraction, it is necessary to examine the constraint of unitary processes. The system exergy represents the maximum amount of work that can be extracted from the system while bringing it into equilibrium with a thermal bath. It can be separated into two parts: ergotropy and residual energy. Thus, the present chapter describes the relationship between exergy and its potential benefits and effects on the performance of quantum batteries.

Keywords: quantum batteries, ergotropy, unitary process, exergy, quantum thermodynamics

1. Introduction

From the late eighteenth century until today, scientists have been experimenting with electrochemical reactions, which allow us to convert chemical energy into electricity. Despite centuries of innovation, the fundamental way that batteries generate energy has remained constant. This process involves an electrochemical reaction that takes place within a closed cell. Over 200 years ago, an Italian scientist, named Alessandro Volta, discovered a reliable electrochemical system. However, Volta's materials and arrangement were inefficient and resulted in a tiny output. Despite this, his work was crucial in laying the foundation for electrochemical advancements. In 1859, Gaston Plante made a significant advancement when he invented a rechargeable battery. Typically, batteries lose their power when their electrodes run out of ions or space to receive them. However, rechargeable batteries are designed so that by applying a charge, the electrodes are either rebuilt or deionized, reversing the primary reaction. In 1859, Gaston Plante made a significant advancement when he invented a rechargeable battery. In 1866, George-Lionel advanced the battery's evolution by creating a closed, portable, and powerful battery based on Volta's original concept. This marked the beginning of the consumer battery era, with Leclanche's invention becoming the first of its kind in the 1890s. His design eventually led to the development of the Duracell alkaline battery in the 1950s. In 1899, Waldemar Jungner made a significant advancement in rechargeable battery technology by combining Leclanche's concept with new developments in materials science. His battery was smaller in size, more durable, and had a higher output than Plante's battery, but had less charge capacity. However, due to its high cost, it did not become commercially viable until the 1930s. Throughout most of the twentieth, it served as a battery technology model, with each type fulfilling its own purpose and meeting technology demands. Battery technology underwent gradual development and finally, in 1985, Akira Yoshino created the first commercially viable rechargeable lithium-ion battery. The introduction of lithium-ion batteries in 1992 revolutionized technology. It meant that complex, energy-intensive gadgets like computers and phones could go mobile, and the technology could be upgraded to energy-consuming levels previously impossible. Furthermore, it largely eliminated the problems associated with the unpredictability of renewable energy production by allowing for industrial-scale energy storage.

Batteries can be defined as devices that store energy from external resources through electrochemical processes and provide that energy to other machines, enabling them to operate remotely without requiring a power resource. In recent years, batteries have become increasingly crucial in terms of size and storage capacity [1]. As devices continue to become smaller, batteries are also shrinking in size. As a result, when their unit cells approach molecules and atoms, quantum mechanical effects must be taken into consideration when describing them [2–4]. Recent theoretical investigations in the field of quantum thermodynamics have demonstrated that entanglement generation is connected to faster work extraction when energy is stored in many-body quantum systems [5]. These findings have prompted research on the use of quantum systems as heat engines and energy storage devices [6]. As a result, there has been a growing interest in the study of quantum batteries, which were first introduced in the influential work by Alicki and Fannes [7]. Scientists are now exploring quantum effects that could potentially enhance the performance of these devices [8–11]. Quantum batteries are theoretical d-dimensional energy storage quantum systems. They are quantum systems with non-degenerate energy levels that use quantum mechanics principles to store energy [7]. Quantum coherence and entanglement are used in the design of these batteries. Quantum batteries have the potential to be more efficient, have a higher energy density, and be smaller and lighter than classical batteries. The most important feature of a good quantum battery is its ability to store as much energy as possible in the shortest time and discharge it optimally. To determine the quality of a quantum battery, one can examine its internal energy and the amount of work that can be extracted from it [12]. Overall, quantum batteries hold great potential for advancing energy storage technology. Furthermore, progress has been made toward implementing experiments [13–17].

In the study of quantum batteries, ergotropy is a critical measure that indicates the amount of energy that can be extracted from a given quantum battery state through the cyclic unitary evolutions [12]. The ergotropy of a battery can change from zero, which represents the passive state [18], to a maximum value calculated from the energy levels of the Hamiltonian and the eigenvalues of the battery's density matrix, as the battery stores or releases energy. Traditionally, quantum mechanics deals with isolated systems that are completely detached from their surrounding environment. The idea of open quantum systems stems from the way physical systems interact with their surroundings. These interactions can cause the transfer of energy, information, or particles between the system and its environment. A major obstacle in studying open quantum systems is decoherence. In quantum physics, decoherence refers to the loss of quantum coherence that arises from the interaction of the quantum system with its environment. Studying quantum batteries from an open quantum systems perspective is crucial as environmental effects on quantum systems are inevitable. There has been extensive research on the impact of environmental parameters on the charging and discharging processes of quantum batteries [19–24]. When external thermal baths couple with the quantum battery, non-unitary effects may occur, causing energy loss due to thermal effects on the system. In some cases, ergotropy can be stored, but a part of the total energy available in the system cannot be stored as ergotropy. Consequently, part of the total energy cannot be extracted from cyclic unitary evolutions. Therefore, it is important to consider the amount of residual energy that cannot be extracted as useful work from open quantum batteries by unitary processes. To better understand the amount of energy lost during work extraction, it is necessary to examine the constraint of unitary processes. The system exergy represents the maximum amount of work that can be extracted from the system while bringing it into equilibrium with a thermal bath. It can be separated into two parts: ergotropy and residual energy [25]. The residual energy cannot be extracted through a unitary process, and it refers to the non-optimal performance of a cyclic thermodynamics process used to extract work from quantum batteries. Thus, the present chapter describes the relationship between exergy and its potential benefits and effects on the performance of quantum batteries.

2. Energy storage and unitary work extraction

A quantum battery is a type of energy storage device that can be represented by a d-dimensional system with a corresponding Hamiltonian as follows

$$\hat{H}_0 = \sum_n \varepsilon_n |\varepsilon_n\rangle\langle\varepsilon_n| \tag{1}$$

with non-degenerate energy levels such that $\varepsilon_n < \varepsilon_{n+1}$. In quantum realm, energy can be stored in both the energy levels and coherence of a state. Therefore, any quantum system with distinct and practically accessible energy eigenstates can be considered as a quantum battery. Charging a quantum battery involves transitioning it from a low- ρ to a high-energy state ρ'. Conversely, utilizing the battery will result in a reduction of its stored energy (see **Figure 1**). In order to understand the charging and extraction of work, a time-dependent control parameter is used to regulate the interaction that occurs during the process. This creates a time-dependent Hamiltonian, $\hat{H}(t) = \hat{H}_0 + V(t)$, which describes the process. To determine the maximum amount of energy that can be extracted from a given system, one must assess a discharging process that begins at $t = 0$ and concludes at $t = \tau$ when the battery is completely drained. It is possible to limit both the charging and discharging processes to be cyclic and unitary. In a unitary process, the energy change must always be considered as a

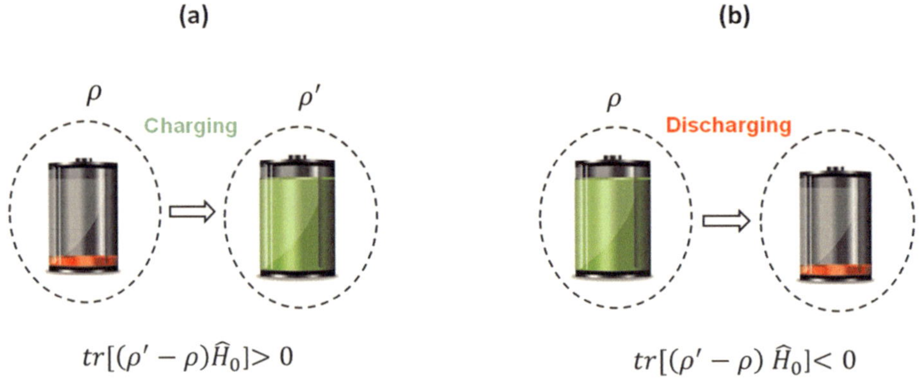

Figure 1.
The time-dependent (a) charging and (b) discharging protocols facilitate the charging and discharging of the battery system through the operation of the Hamiltonian H, which is activated for the duration of the time interval $(0, \tau)$.

work. The terms "work storing" and "battery charging" are used interchangeably, as are "work extraction" and "battery discharging" when using the battery. The average extracted work is given by the system's unitary operator generated by the full Hamiltonian as follows

$$W = tr\left[\hat{H}_0\rho_0\right] - tr\left[\hat{H}_0\hat{U}\rho_0\hat{U}^\dagger\right] \tag{2}$$

where ρ_0 refers to the initial state of the system. The maximum amount of work that can, on average, be extracted from the system is given by:

$$\mathcal{E} = tr\left[\hat{H}_0\rho_0\right] - \min_{\hat{U}\in SU(d)} tr\left[\hat{H}_0\hat{U}\rho_0\hat{U}^\dagger\right] \tag{3}$$

Then, the energy stored in a quantum battery when it's in state ρ_0 can be calculated by finding the difference between the system's internal energy $tr\left[\hat{H}_0\rho_0\right]$ and the energy $tr\left[\hat{H}_0\varrho_\rho\right]$ of the lowest accessible state ϱ_ρ of system, which is passive by definition [18]. Accordingly, a state ϱ is passive if $tr[H_0\varrho] \le tr\left[H_0 U\varrho U^\dagger\right]$ for all unitaries U, or, equivalently, if and only if $\varrho = \sum_{n=1}^{d} r_n|\varepsilon_n\rangle\langle\varepsilon_n|$ for $r_n \ge r_{n+1}$, i.e., it commutes with the internal Hamiltonian H_0 and has non-increasing eigenvalues. This difference is known as *ergotropy* [12].

$$\mathcal{E} = tr\left[\hat{H}_0\rho_0\right] - tr\left[\hat{H}_0\varrho_\rho\right] \tag{4}$$

A practical and intuitive method for constraining the extractable work of Eq. (4) is to examine a thermal state with equivalent entropy to passive state ϱ_ρ, while also minimizing the energy in respect to H_0. It has been demonstrated that the lower bound of ergotropy for an arbitrary state ρ is expressed as:

$$\mathcal{E}_{\max} \le tr\left[\hat{H}_0\rho\right] - tr\left[\hat{H}_0\rho_{\overline{\beta}}\right] \tag{5}$$

where the canonical Gibbs state denoted by $\rho_\beta = \exp\left[-\beta\hat{H}_0\right]/tr\left(\exp\left[-\beta\hat{H}_0\right]\right)$, with an inverse temperature β, and $\overline{\beta}$ is selected to ensure that the von Neumann entropy $S(\rho) = -tr[\rho \ln \rho]$ of state ρ is equal to that of state $\rho_{\overline{\beta}}$ [7]. All thermal states exhibit passivity, and in the context of two-level systems, every passive state can be characterized as thermal, as it is always possible to define a (positive or negative) temperature.

The usual scenarios involve the interaction of the quantum system with the external environment, resulting in decoherence and the depletion of quantum resources, thus defining it as an open quantum system. An open quantum battery refers to an open quantum system capable of interacting with the external environment, whether with or without a mediator. As explained earlier, if the battery system is kept isolated, closed, or protected from any external influences, it will always evolve in a unitary manner as described. However, in the case of an open quantum battery, its dynamics are determined by the global Hamiltonian as follows:

$$H = H_0 + H_C + H_E + H_{\text{int}} \tag{6}$$

where H_{int} represents the interaction between the battery and/or charger and the environment. The dynamics of an open battery, described by the Hamiltonian of Eq. (6), can be expressed using quantum master equations [26]. This dynamic has been extensively studied under the Markovian regime, employing the quantum master equation in Lindblad form as follows [26]

$$\dot{\rho}_0(t) = -i\left[\hat{H}_0, \rho(t)\right] + \mathcal{D}[\rho(t)] \tag{7}$$

The super-operator $\mathcal{D}[\rho(t)] = \sum_k \gamma_k \left(\hat{L}_k \rho \hat{L}_k^\dagger - \frac{1}{2}\left\{\hat{L}_k^\dagger \hat{L}_k, \rho\right\}\right)$ models the battery's interaction with the universe via operators $\hat{L}_k$ that cause incoherent transitions at rates γ_k. Some research have also explored open quantum batteries beyond the Markovian approximation [19, 20]. In non-Markovian regime, the inclusion of memory effects arising from the interaction with the environment and the charger is crucial for a comprehensive depiction of the battery's temporal evolution.

The implementation of a quantum battery in practice must address the challenge of environmental interactions. As a result, safeguarding against energy leakages and decoherence is crucial for the successful realization of such devices. As a result of these interactions, the entropy level of the battery increases, making unitary evolution insufficient to rectify entropy production and stabilize the system. Notably, the presence of decoherence effects during the charging process negatively impacts the performance of operational quantum batteries, leading to the self-discharge phenomenon [27]. Efforts have been made to mitigate the quantum battery's interactions with the environment to prevent eventual deactivation, yet certain approaches aim to transform the environment's role from detrimental to beneficial. Where non-unitary discharging provides more charge due to availability or exergy compared to a cyclic unitary process [25].

3. Exergy in a quantum battery

When the system's temperature deviates from that of the surrounding environment, it fails to achieve mutual stable equilibrium. Consequently, the requisite conditions for mutual thermal equilibrium are not fulfilled. Any absence of mutual stable equilibrium between a system and its environment can be leveraged to generate work.

In this section, consider work extraction from a quantum system with Hamiltonian H in an arbitrary non-equilibrium state ρ. In the most straightforward scenario, solely unitary operations are permitted, wherein the Hamiltonian undergoes change following a certain cyclic protocol, such that H retains its initial value both before and after the operation. The framework's natural extension could involve the incorporation of non-unitary transformations to enhance performance. Specifically, if non-unitary transformations are added alongside unitary cyclic driving and contact with a thermal reservoir, the extracted work may increase as a result of both the system entropy changing during the protocol and the reservoir supplying additional energy. In such a scenario, the maximum amount of extractable work is determined by the disparity in non-equilibrium free energy between the state ρ and the thermal equilibrium state at the reservoir inverse temperature $\beta(k_B = 1)$ as follows

$$\mathcal{W}_{ext} = \mathcal{F}(\rho) - \mathcal{F}(\rho_\beta) = \beta^{-1} S(\rho \| \rho_\beta) \tag{8}$$

where $S(\rho\|\sigma) = tr[\rho(\ln \rho - \ln \sigma)]$ stands for the quantum relative entropy [28]. Implementing an operationally reversible isothermal process enables the attainment of this outcome [29, 30]. The non-equilibrium free energy for a system in state ρ, with Hamiltonian H, and with respect to a thermal bath at reverse temperature β is defined as

$$\mathcal{F}(\rho) = tr[H\rho] - \beta^{-1} S(\rho) \tag{9}$$

The quantity W_{ext} represents the extractable work from a system as it reaches equilibrium with a thermal reservoir, which can be referred to as *exergy* $\Sigma^{\rho \to \rho_\beta}$ in analogy with the classical definition of exergy in thermodynamics by Zoran Rant [31], originating from the Greek term "ex" [$\varepsilon\xi$] and "ergon" [$\varepsilon\rho\gamma o\upsilon$]. Evidently, given that Eq. (8) exergy is the information gain, up to a factor β and it is considered a fundamental quantity both in statistics and physics. In fact, the exergy of a system is characterized as the utmost work achievable by the combination of the system and a defined thermal bath. The bath is presumed to be infinite, in equilibrium, and encompassing all other systems. Generally, the bath is determined by specifying its temperature, pressure, and chemical composition. Exergy is not solely a thermodynamic property, but rather a property of both a system and the bath.

In other instances, it should be noted that the system may be considered to be in contact with a reservoir that adheres to fermionic (or bosonic) statistics with a chemical potential of μ and a temperature of T. Within this context, the system also has the ability to exchange particles with the reservoir. Where the reservoir is in own local thermal equilibrium state $\rho_\beta = Z^{-1} \exp\left[-\beta\left(\hat{H} - \mu\hat{N}\right)\right]$ with partition function $Z = tr\left(\exp\left[-\beta\left(\hat{H} - \mu\hat{N}\right)\right]\right)$ and particle number operator $\hat{N}$. In this way, the maximum available work or exergy is delineated as follows [32].

$$\Sigma^{\rho \to \rho_\beta} = \Lambda(\rho) - \Lambda(\rho_\beta) = \beta^{-1} S(\rho \| \rho_\beta) \tag{10}$$

where $\Lambda(\hat{\rho}) = E - \mu N - TS(\rho)$ is the non-equilibrium grand potential with $E = tr\left(\rho \, \hat{H}\right)$, and $N = tr\left(\rho \, \hat{N}\right)$ being the energy, and particle number of the system, respectively.

3.1 Balance equation

After the extraction of ergotropy, it is established that the internal energy of the system does not diminish to zero [12, 33], thereby leaving a residual amount of energy still accessible within the system. To measure the amount of energy that can be extracted following ergotropy's extraction, one can analyze the system's dynamics, as illustrated in **Figure 2** [25]. The available work in the initial state ρ_0 can be extracted in two distinct manners: (1) through a unitary process that transitions the system into a passive state ε_{ρ_0}, thereby quantifying the extractable work as ergotropy, and (2) via a non-unitary process that guides the system to the thermal equilibrium state ρ_β, quantifying the extractable work as the variation in free energy. As depicted in **Figure 2**, it becomes feasible to identify a specific amount of energy denoted as $\Sigma(\varepsilon_{\rho_0} \to \rho_\beta)$, such that the exergy $\Sigma(\rho_0 \to \rho_\beta)$ is extracted in process $\rho_0 \to \rho_\beta$, and the extractable ergotropy $\mathcal{E}$ in process $\rho_0 \to \varepsilon_{\rho_0}$ adheres to the balance equation [25]

$$\Sigma^{\rho_0 \to \rho_\beta} = \mathcal{E}(\rho_0) + \Sigma^{\varrho_{\rho_0} \to \rho_\beta} \tag{11}$$

where $\Sigma^{\varrho_{\rho_0} \to \rho_\beta} = \left[\mathcal{F}\left(\varrho_{\rho_0}\right) - \mathcal{F}(\rho_\beta)\right]$ is the exergy of passive state ϱ_{ρ_0} and represents an available work which unitary processes are unable to extract. In view of these factors, it can be demonstrated that the reduction in free energy exceeds the ergotropy, i.e., $\Sigma^{\rho_0 \to \rho_\beta} \geq \mathcal{E}(\rho_0)$, where equality is attained only when passive state exergy vanishes. It is important to observe that the existence of the environment serves a beneficial purpose, providing an additional energy source and enhancing our capacity to harness work.

This outcome has dual interpretations. The initial interpretation concerns the distinctive nature of an energetically efficient starting point for storing ergotropy. When a system interacts with a reservoir, the initial ergotropy is stored in a non-pure state ρ_0. Then, through an optimal unitary operation $U^\star$, the system can be driven for a brief period to extract ergotropy. By appropriately selecting the initial state so that $U^\star \rho_0 U^{\star\dagger} \to \rho_\beta$, it becomes apparent that $\Sigma^{\varrho_{\rho_0} \to \rho_\beta} = 0$, resulting in $\Sigma^{\rho_0 \to \rho_\beta} = \mathcal{E}(\rho_0)$. In summary, all the available energy of the system can be harnessed as ergotropy. Consequently, given the uniqueness of the thermal state ρ_β, the optimal unitary operation $U^\star$ leads to the uniqueness of ρ_0. The second scenario involves the efficient extraction of energy. Moreover, since the quantity $\Sigma^{\varrho_{\rho_0} \to \rho_\beta}$ cannot be negative, the energy lost during ergotropy extraction is an anticipated natural outcome owing to entropy production. The second scenario can be seen as a direct application of the second law of thermodynamics to quantum batteries. Given that $\Sigma^{\varrho_{\rho_0} \to \rho_\beta} > 0$, one can write

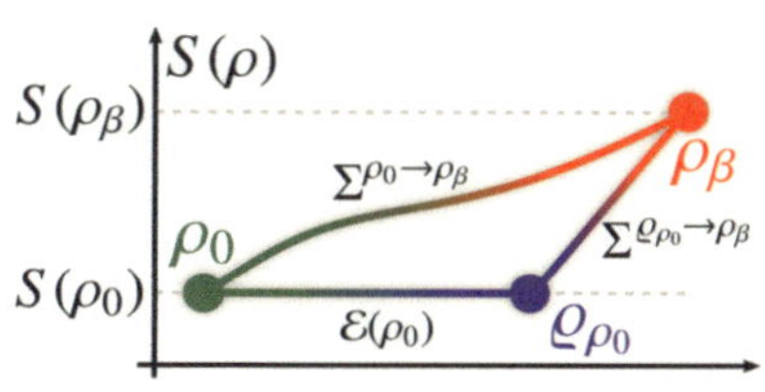

Figure 2.
A visual illustration of the three states under discussion is depicted in the schematic. The extraction of ergotropy, which leaves the system entropy unchanged, is succeeded by the thermalization process, leading the system to reach equilibrium with a thermal reservoir at reverse temperature β [25].

$$\Delta S > \beta\left(tr\{H\rho_\beta\} - tr\left\{H\varrho_{\rho_0}\right\}\right) \tag{12}$$

Here, $\Delta S = S(\rho_\beta) - S(\varrho_{\rho_0})$ represents the amount of entropy production necessary to extract the stored exergy within the system. Considering that the heat exchanged during thermalization between the system and reservoir is determined by the internal energy variation of the system [34], it can be inferred that $\Delta S > \beta Q$.

4. Conclusions

This chapter addresses the dissipation of energy during cyclic unitary work extraction from a quantum system, with a focus on the identification of this energy loss as the exergy of the quantum passive state. It was demonstrated that, in a real-world scenario, the ergotropy results in energy loss due to the constraints of the unitary process. Additionally, the presence and uniqueness of an optimal passive state for ergotropy and exergy extraction is examined by taking into account the system-bath interaction that leads to thermalization. With respect to the second law of thermodynamics, our key finding was explained as a natural outcome of the entropy production during the thermalization process for exergy extraction. Our results also led to the identification of a range of ergotropy and exergy extraction processes where the total quantum correlations of the system (measured through quantum discord) are conserved. This implies that the exergy of a quantum passive state can be preserved as quantum correlations. As exergy represents the amount of energy recoverable through a thermalization process, our findings open up new possibilities for developing operational protocols for open quantum batteries.

Acknowledgements

This work has been supported by the University of Kurdistan. F.H. Kamin and S. Salimi thank Vice Chancellorship of Research and Technology, University of Kurdistan.

Conflict of interest

The authors declare no conflict of interest.

Exergy and Quantum Batteries
DOI: http://dx.doi.org/10.5772/intechopen.1004642

Author details

Farzaneh Hatami Kamin* and Shahriar Salimi
Department of Physics, University of Kurdistan, Sanandaj, Iran

*Address all correspondence to: f.hatami@uok.ac.ir

References

[1] Scrosati B. Modern Batteries: An Introduction to Electrochemical Power Sources. 2nd ed. London: Arnold; 1997. p. 351

[2] Campaioli F, Pollock FA. Vinjanampathy S. In: Binder F, Correa L, Gogolin C, Anders J, Adesso G, editors. Thermodynamics in the Quantum Regime: Fundamental Aspects and New Directions. Cham: Springer; 2018. pp. 207-225. DOI: 10.1007/978-3-319-99046-0_8

[3] Goold J, Huber M, Riera A, del Rio L, Skrzypczyk P. The role of quantum information in thermodynamics-a topical review. Journal of Physics A: Mathematical and Theoretical. 2016;**49**: 143001. DOI: 10.1088/1751-8113/49/14/143001

[4] Vinjanampathy S, Anders J. Quantum thermodynamics. Contemporary Physics. 2016;**57**:545-579. DOI: 10.1080/00107514.2016.1201896

[5] Hovhannisyan KV, Perarnau-Llobet M, Huber M, Acín A. Entanglement generation is not necessary for optimal work extraction. Physical Review Letters. 2013;**111**:240401. DOI: 10.1103/PhysRevLett.111.240401

[6] Levy UA, Kosloff R. Quantum heat machines equivalence, work extraction beyond Markovianity, and strong coupling via heat exchangers. Entropy. 2016;**18**:124. DOI: 10.3390/e18040124

[7] Alicki R, Fannes M. Entanglement boost for extractable work from ensembles of quantum batteries. Physical Review E. 2013;**87**:042123. DOI: 10.1103/PhysRevE.87.042123

[8] Campaioli F, Pollock FA, Binder FC, Céleri L, Goold J, Vinjanampathy S, et al. Enhancing the charging power of quantum batteries. Physical Review Letters. 2017;**118**:150601. DOI: 10.1103/PhysRevLett.118.150601

[9] Binder FC, Vinjanampathy S, Modi K, Goold J. Quantacell: Powerful charging of quantum batteries. New Journal of Physics. 2015;**17**:075015. DOI: 10.1088/1367-2630/17/7/075015

[10] Gyhm J-Y, Šafránek D, Rosa D. Quantum charging advantage cannot Be extensive without global operations. Physical Review Letters. 2022;**128**: 140501. DOI: 10.1103/PhysRevLett.128.140501

[11] Kamin FH, Tabesh FT, Salimi S, Santos AC. Entanglement, coherence, and charging process of quantum batteries. Physical Review. 2020;**102**: 052109. DOI: doi.org/10.1103/PhysRevE.102.052109

[12] Allahverdyan AE, Balian R, Nieuwenhuizen TM. Maximal work extraction from finite quantum systems. Europhysics Letters. 2004;**67**:565. DOI: 10.1209/epl/i2004-10101-2

[13] Quach JQ, McGhee KE, Ganzer L, Rouse DM, Lovett BW, Gauger EM, et al. Superabsorption in an organic microcavity: Toward a quantum battery. Science Advances. 2022;**8**:eabk3160. DOI: 10.1126/sciadv.abk3160

[14] Hu C-K, Qiu J, Souza PJP, Yuan J, Zhou Y, Zhang L, et al. Optimal charging of a superconducting quantum battery. Quantum Science and Technology. 2022; 7:045018. DOI: 10.1088/2058-9565/ac8444

[15] Joshi J, Mahesh TS. Experimental investigation of a quantum battery using star-topology NMR spin systems.

Physical Review A. 2022;**106**:042601. DOI: 10.1103/PhysRevA.106.042601

[16] Gemme G, Grossi M, Ferraro D, Vallecorsa S, Sassetti M. IBM quantum platforms: A quantum battery perspective. Batteries. 2022;**8**:2313-0105. DOI: 10.3390/batteries8050043

[17] Huang X, Wang K, Xiao L, Gao L, Lin H, Xue P. Demonstration of the charging progress of quantum batteries. Physical Review A. 2023;**107**:L030201. DOI: 10.1103/PhysRevA.107.L030201

[18] Pusz W, Woronowicz SL. Passive states and KMS states for general quantum systems. Communications in Mathematical Physics. 1978;**58**:273-290. DOI: 10.1007/BF01614224

[19] Ghosh S, Chanda T, Mal S, Sen(De) A. Fast charging of a quantum battery assisted by noise. Physical Review A. 2021;**104**:032207. DOI: 10.1103/PhysRevA.104.032207

[20] Kamin FH, Tabesh FT, Salimi S, Kheirandish F, Santos AC. Non-Markovian effects on charging and self-discharging process of quantum batteries. New Journal of Physics. 2020;**22**:083007. DOI: 10.1088/1367-2630/ab9ee2

[21] Kamin FH, Abuali Z, Ness H, Salimi S. Quantum battery charging by non-equilibrium steady-state currents. Journal of Physics A: Mathematical and Theoretical. 2023;**56**:275302. DOI: 10.1088/1751-8121/acdb11

[22] Barra F. Dissipative charging of a quantum battery. Physical Review Letters. 2019;**122**:210601. DOI: 10.1103/PhysRevLett.122.210601

[23] Pirmoradian F, Mølmer K. Aging of a quantum battery. Physical Review A. 2019;**100**:043833. DOI: 10.1103/PhysRevA.100.043833

[24] Tabesh FT, Kamin FH, Salimi S. Environment-mediated charging process of quantum batteries. Physical Review A. 2020;**102**:052223. DOI: 10.1103/PhysRevA.102.052223

[25] Kamin FH, Salimi S, Santos AC. Exergy of passive states: Waste energy after ergotropy extraction. Physical Review E. 2021;**104**:034134. DOI: 10.1103/PhysRevE.104.034134

[26] Breuer HP, Petruccione F. The Theory of Open Quantum Systems. New York: Oxford University Press; 2007. 625 p. DOI: 10.1093/acprof:oso/9780199213900.001.0001

[27] Santos AC, Çakmak B, Campbell S, Zinner NT. Stable adiabatic quantum batteries. Physical Review E. 2019;**100**: 032107. DOI: 10.1103/PhysRevE.100.032107

[28] Vedral V. The role of relative entropy in quantum information theory. Reviews of Modern Physics. 2002;**74**:197. DOI: 10.1103/RevModPhys.74.197

[29] Skrzypczyk P, Short AJ, Popescu S. Work extraction and thermodynamics for individual quantum systems. Nature Communications. 2014;**5**:4185. DOI: 10.1038/ncomms5185

[30] Parrondo JMR, Horowitz JM, Sagawa T. Thermodynamics of information. Nature Physics. 2015;**11**: 131-139. DOI: 10.1038/nphys 3230

[31] Rant Z. Ein neues Wort fur technische Arbeits-fahigkeit. Forschung im Ingenieurwesen. 1956;**22**:36-37

[32] Schlögl F. Probability and Heat: Fundamentals of Thermostatistics. Wiesbaden: Vieweg+Teubner Verlag Springer; 2013. p. 252

[33] Moraes LFC, Saguia A, Santos AC, Sarandy M. Charging power and stability of always-on transitionless driven quantum batteries. EPL. 2021;**136**:23001. DOI: 10.1209/0295-5075/ac1363

[34] Alicki R. The quantum open system as a model of the heat engine. Journal of Physics A: Mathematical and General. 1979;**12**:L103. DOI: 10.1088/0305-4470/12/5/007

Section 3

Exergy Applied to Liquid and/or Fluid

Chapter 5

Heat Transfer Fluids and External Convection Effects on Transient Thermodynamic Behavior of a Parabolic Trough Solar Collector in Sahelian Climate

Steven Audrey Heugang Ndjanda and Etienne Tchoffo Houdji

Abstract

This chapter deals with the comparison of four heat transfer fluids (HTFs) for enhanced performance in heat transfer rates on the Sandia experimental LS-2 collector in Sahelian climate. The tested HTFs are dry air, liquid water, Therminol VP-1™, and Syltherm 800. In the literature, the thermal modeling of the parabolic trough solar collectors (PTSCs) is done by assuming: the steady-state or quasi-steady-state regime, the constant properties of HTFs, and others. These assumptions underestimate the transient heat transfer through the HTF flow. Moreover, the best theoretical performance of PTSCs evaluated in terms of idealized processes has not been developed satisfactorily yet. Subsequently, none effective assessment method of the PTSC has been experienced to investigate its transient thermal and thermodynamic sensitivity to (1) the collector, (2) operational, and (3) meteorological parameters. The implemented numerical approach here follows the transient analysis on the first and second laws of thermodynamics published recently. The physical model of a tubular receiver comprised an outer glass envelop, absorber pipe, and heat transfer fluid. Their respective heat rate balance equations were reliably established, discretized by the finite volume method, and efficiently computed. Then, a transient exergy-degradation analysis is conducted and validated using the available experimental results.

Keywords: parabolic trough solar collector, heat transfer fluids, specific mass flow rate, wind flow, numerical simulations, Sahelian climate

1. Introduction

Solar concentration technologies have gained considerable attention on design, theoretical and practical views and can be applied to generate cost-effective electricity. Thermal solar electricity is generated by concentrating solar energy using a concentrating solar power (CSP), by heating fluid and producing steam, which is then used to power an electrical generator. The parabolic trough solar collectors (PTSC) are the most proven widespread and commercially tested solar energy technologies available for harnessing solar resources [1–6].

PTSC is essentially composed by the highly reflective mirrors and a receiver tube—also called the heat collection element (HCE) placed at the focal line of the cylindrical concentrator. The receiver tube consists of an absorber pipe enclosed in a glass envelop. The absorber transfers the incident solar concentrated radiation reflected by the mirrors to a heat transfer fluid (HTF) circulating through the absorber. At normal operating conditions, HTF departs from a collector with a certain outlet temperature. The main task of the HTF is to accumulate the thermal energy from the solar field and to transport it into the storage or power block. In many studies, gases (e.g. dry air, steam), liquid water, thermal oils (e.g. Therminol VP-1™, Dowtherm A), nanofluids (e.g. CuO-Syltherm 800, Ag–MgO with water), molten salts (e.g. Hitec, Hitec XL), liquid metals (e.g. sodium-potassium alloys: NaK, pure Lead-Bismuth Eutectic: LBE) and vegetable oils (e.g. Rapeseed, Jatropha) are various types of HTFs currently used. Heat exchangers are employed to transfer energy to steam in a power plant. A steam turbine is used as the prime mover for power generation.

It is well-known that the type of HTF can enhance the PTSC's performance by significantly affecting the power cycle efficiency and the system cost. There are some expected properties from HTFs. Also, according to Montes et al. [7], the evaluation of HTFs considered in PTSCs, must include the following parameters: corrosion and durability, control strategies during daily operation with startup, shutdown, and sudden changes in weather conditions, among others. Let consider particularly the handling conditions that constantly act on the collectors from startup to shutdown of a daily operation. Thus, PTSCs always have to operate under transient considerations, especially with HTFs involving thermophysical properties that are strongly responsive to working temperature changes.

Furthermore, analyzing HTFs effects on PTSC implies the consideration of (a) the type of the flow, (b) the inlet temperature of fluid, (c) the mass flow rate [8], and (d) the type of HTF, as the sensitive parameters [9].

The reviewed literature shows that a lot of studies have been carried out to model the performance of parabolic trough collectors. In this regard, Montes et al. [7] compared the performance of the parabolic trough collectors using oil, water/steam, or molten salt as heat transfer fluids. Ouagued et al. [10] developed a one-dimensional model dividing the HCE into several segments for PTCs. The numerical results showed that most of the heat gained was collected in the south of Algeria, and the best thermal oil is Syltherm 800. Sa et al. [11] have proposed a transient modeling of PTSC; they used molten salt as an HTF, which replaces the conventional thermal oil. The maximum temperature achieved was 520°C, and the average minimum temperature was 352°C. Marif et al. [12] presented the optical and thermal performance of a PTSC under the climatic conditions of the Algerian Sahara. Two fluids were considered, liquid water and TherminolVP-1™ synthetic oil. Hussein et al. [13] have compiled a detailed study of recent advances with the application of nanotechnology in different kinds of solar collectors. Ghoneim et al. [14] proposed a numerical model to study the effect of different collector parameters and operating conditions on the parabolic trough solar collector performance in the Kuwait climate. Najjar and Sadeq [15], with parabolic trough collector's system, have presented the performance of a supercritical organic Rankine cycle in a typical environment of Abu Dhabi, United Arab Emirates. They used R500 as an organic working fluid, and the heat transfer fluid used was the Therminol VP-1. Wang et al. [16] designed, fabricated, and tested on a pony parabolic concentrator, a cost-effective single-pass all-glass PTR, consisting of double glass tubes with different diameters, bellows, and stainless steel connecting tubes on both ends. Kessel et al. [17] have proposed an evaluation of the thermal performances of a PTSC under the climatic conditions of the Cameroon Sahelian zones using dry air, liquid water, and TherminolVP-1™ as HTFs. However, no information concerning the transient thermodynamic behavior of PTSC is provided; also, the effects of mass flow rate and wind flow are not studied. Yılmaz and Mwesigye [18] have proposed a reviewed study to improve the thermal performance of PTSCs by modifying and improving the thermophysical properties of the conventional HTF used. Okonkwo et al. [19] carried out a sensitivity analysis to understand the heat transfer parameters that could enhance the thermal performance of the LS-2 collector. Six working fluids were investigated: pressurized water, supercritical carbon dioxide (CO_2), Therminol VP-1™, and the addition of copper oxide (CuO), iron (II, III) oxide or magnetite (Fe_3O_4), and aluminum oxide (Al_2O_3) nanoparticles to TherminolVP-1™, and the Al_2O_3-Oil nanofluid provides the best improvement of 0.22% to thermal efficiency. Bellos et al. [20] investigated six different nanoparticles dissolved in hot oil. Effect of varying concentration ratio, solar intensity, and flow rate were calculated using Engineering Equation Solver. Bhargav et al. [8] presented a detailed review of work done optimizing of geometrical components for various applications of parabolic trough collector. Evangelisti et al. [21] investigated a review of the latest advances on solar thermal collectors. They have presented six heat transfer mediums: water, glycol water, molten salts, hydrocarbon oils, phase change material, and nanofluids. Okonkwo et al. [22] proposed a validated model using the Sandia National Laboratory, AZTRAK, platform results. The absorber tube is modeled for five different geometry configurations: a plain tube (smooth tube), a longitudinal finned tube, a tube with a porous insert, a tube with a twisted tape insert, and a wavy (converging-diverging) tube. Al-Oran et al. [23] conducted a study aiming to improve LS-2 parabolic trough model and compare the enhancement effect using different mono and hybrid nanofluids. Inserting mono nanoparticles of aluminum oxide (Al_2O_3), cerium oxide (CeO_2), copper oxide (CuO), and hybrid combinations of Al_2O_3 with CeO_2 or CuO nanoparticles in a Syltherm 800 was investigated by five different cases. Abed et al. [24] conducted a parametric comparison between pure fluids, considering the effect of various inlet fluid temperatures and different Reynolds (Re) numbers on thermal performance. They assessed three main types of pure fluids: water, Therminol VP-1™, and molten salt. Zaharil et al. [25] in their research aimed to evaluate and compare the performances of six different HTFs (Pressurized water, Syltherm 800, Therminol VP-1™, Solar salt, Hitec XL, Liquid sodium) under a wide range of climatic conditions by using the Malaysian climatic condition as reference. Mwesigye and Yilmaz [26] investigated numerically the thermal and thermodynamic performance of a PTSC system with a geometrical concentration ratio valuing 113. More recently, El Kouche et al. [27] modeled the thermal behavior of a parabolic trough collector where the heat transfer fluid has temperature-dependent physical properties.

To summarize, it can be noted that a realistic unsteady state heat exchange model of the PTSC, accounting the variable thermophysical properties of the HTF with the temperature has never been completely proposed in the literature. To account the temperature-dependent variable thermophysical properties of the considered HTFs and all thermal mechanisms, which have been neglected in existing models, a more reliably heat-transfer analysis is carried out in the present research. Also, the second law of the thermodynamic unsteady analysis or the transient exergy-destruction analysis has not extensively been applied on the PTSC according to available literature.

Furthermore, three main groups of critical parameters were investigated in the analysis of the thermal and thermodynamic behavior of the parabolic trough solar collector:

a. The first group of parameters (also called the collector parameters) is composed by the type of the HTF, the solar tracking system, type of reflectors, the concentration ratio of the collector, cross section of the glass envelop and absorber pipe, their length and respectively constitutive material, and the presence or not of the annulus flow.

b. The second group of parameters (also called the operational parameters) is composed by the inlet temperature, the mass rate flow of the fluid, and the type regime of the flow (either hydrodynamic or thermal).

c. The third group of parameters (also called the meteorological parameters) is composed by the ambient temperature, solar irradiance, and wind flow around the tubular receiver.

None effective assessment approach of the PTSC has been at present experienced, to investigate his transient thermal and thermodynamic sensitivity to the collector, operational and meteorological parameters. The present work follows the transient both energy and exergy analysis of a parabolic trough solar collector using temperature-dependent thermophysical properties of the HTFs previously developed, tested, and validated with available experimental data in [9]. Now, a comparative numerical analysis of the thermal and exergy transfers into the Sandia experimental LS-2 type collector in Sahelian climate for three types of HTFs is proposed: gas (dry air), liquid water, and thermal oils (Therminol VP-1™ and Syltherm 800). Appendix A shows the correlation of the thermophysical properties of the HTFs used.

On the other hand, the PTSCs are usually located in an open land where strong winds may occur and affect the heat exchange between the solar receiver tube and the ambient air. Thus, concerning the wind velocity influence on PTSC, many researchers have tried different shapes of torque tube, such as cylindrical hollow tube box-type structures. They have investigated the thermal effect of wind velocity on the receiver tube of PTSCs. Hachicha [28] investigated the wind speed effect on the flow field and heat transfer around a parabolic trough solar collector. In another work, Paetzold et al. [29] investigated the airflow in a parametric study reduction of wind load and thermal sand losses in the receiver tubes using the commercial CFD program ANSYS CFX 14.5. Fu et al. [30] simulated the wind structure interaction and optimized their new model. They reduced the collector weight about 5.8% in comparison with the previous model. Hassane et al. [31] investigated the effect of wind speed on the thermal performance of the parabolic trough solar collector in quasi-steady regime operation in the climatic conditions of the city of N'Djamena located in a Sahelian region of the Chad Republic.

We also aim to investigate the transient thermal sensitivity of the LS-2 with the following parameters: specific mass flow rate of an HTF and the wind velocity around the tubular receiver of PTSC.

This chapter is organized as follows: Model of PTSC and thermodynamic implementation devoted to a brief presentation of our model of PTSC considered, the development of transient thermodynamic implementation, and the numeric analyses, and results and discussion, here are highlighted the effects of the following sensitive parameters considered on the Sandia experimental LS-2 type, under the Sahel climate: type of the HTFs, the specific mass flow rate, and wind speed around the receiver.

2. Model of PTSC and thermodynamic implementation

2.1 Thermal model of PTSC

A schematic of the parabolic trough solar collector system under consideration is shown in **Figure 1** in which the HCE is a layered system of the glass envelop, the absorber pipe, and HTF. The main energy fluxes appearing in the transient heat transfer equations of the different components of the layered system are known. Mathematically, the system can be described by the transient heat transfer equation for: the glass envelop, the absorber pipe, and HTF. The derivation of these equations is obtained by applying the energy balance principle within each layer element. In other words, the heat transfer model is an energy balance between the heat transfer fluid and its surroundings.

Figure 2 shows the thermal network model in a cross section of the HCE. Readers are encouraged to consult the details, modeling of direct solar radiation and solar

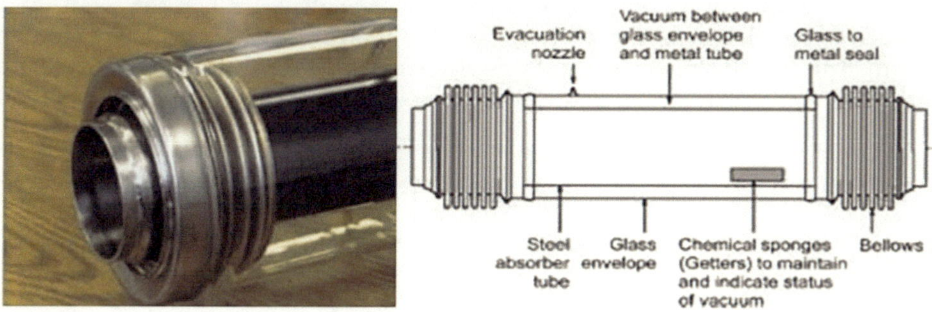

Figure 1.
Heat collection element, adapted from [17]. Useful absorptivity (95%) and useful emissivity (14%) with about 350°C and 400°C supported.

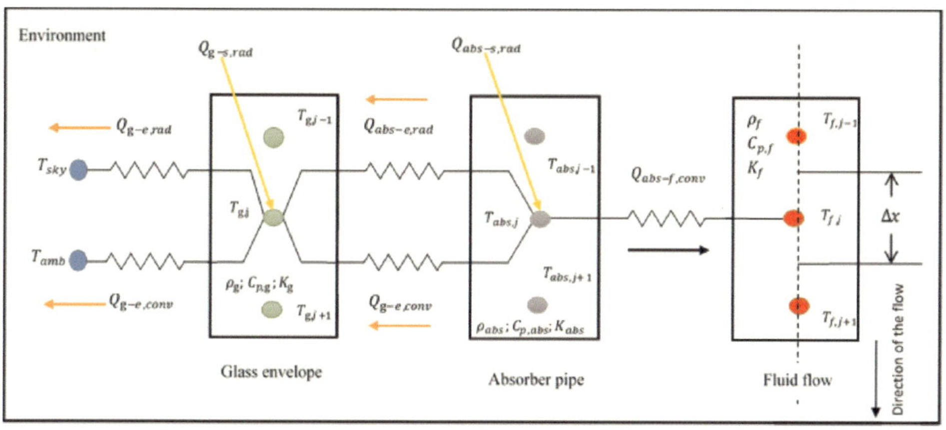

Figure 2.
The proposed thermal network model of parabolic trough solar collector.

radiation availability, assumptions of the heat transfer model, heat transfer balance equations of the heat collector element, and the numerical strategies developed in [9]. It is, therefore, considered that the equations of the mathematical model of the transient heat transfer of glass envelop, absorber pipe, and heat transfer fluid are solved in the cases of the problems considered below, and thus temperatures of these subsystems are simulated. The heat transfer model was compared with other solar receiver models [32]. The obtained temperatures of glass envelop absorber pipe and HTF are then used to build the following energy and exergy analysis.

2.2 Unsteady state energetic analysis

By definition, the transient collector thermal efficiency is the ratio of the useful flux delivered to the incident solar irradiance on the concentrator aperture

$$\eta_{th}(t) = \frac{\dot{Q}_{uf}}{A \times I_d} \tag{1}$$

where the useful flux is given as $\dot{Q}_{uf}$ defined by Eq. (1), by unit of length. Net useful flux or heat gained can be calculated as the total convection heat transfers from the inner surface of the absorber tube to the HTF. The term $A \times I_d$ is the input irradiance energy, while the quantity $A = a$ is the aperture area of the collector by unit of length. The overall daily average of this efficiency is given as presented in [9, 33] as follows:

$$\eta_{th,d} = \int_{t_{sunrise}}^{t_{sunset}} \dot{Q}_{uf} dt \Big/ \int_{t_{sunrise}}^{t_{sunset}} (A \times I_d)\, dt \tag{2}$$

where $t_{sunrise}$ and t_{sunset} are respectively the computed sunrise and sunset times of the considered day.

2.3 Transient exergetic analysis

Padilla et al. [34] and Heugang et al. [9] give the exergy gained by the HTF due to the incident irradiance on the solar collector as below:

$$\dot{E}_{ex_{gain,f}} = \dot{m}_f \left\{ \int_{T_{in}}^{T_{out}} C_{P,f}(T) dT - T_{amb} \int_{T_{in}}^{T_{out}} C_{P,f}(T) \frac{dT}{T} - \frac{\Delta P}{\rho_f} + \frac{\Delta V_f^2}{2} \right\} \tag{3}$$

In order to be able to compare present results with results obtained in the literature, for installations of different sizes, specific mass flow rates (mass flow rate per the inner absorber pipe cross-sectional area: $\dot{m}_f$) are expressed in kg/s.m.^{2}

The first two terms of the Eq. (3) above represent respectively the exergy gained due to an increase in the HTF temperature due to the solar insolation and the flow friction. The last terms represented in this equation correspond to a reduction of the mechanical energy due to flow friction and the increase of exergy due to kinetic energy, respectively. Considering the Carnot efficiency expression between the local fluid temperature T and the ambient temperature T_{amb} as formulated by Moran et al. [35]:

$$\eta_{Carnot} = (1 - T_{amb}/T) \tag{4}$$

Thus, as the authors suggested in a previous published work, Eq. (3) can be rewritten in terms of Carnot efficiency, as follows [9]:

$$\dot{E}_{ex_{gain,f}} = \dot{m}_f \left\{ \int_{T_{in}}^{T_{out}} (1 - T_{amb}/T) C_{P,f}(T) dT - \frac{\Delta P}{\rho_f} + \frac{\Delta V_f^2}{2} \right\} \tag{5}$$

Fujiwara [36] proposed the expression of the pressure drop appearing in Eq. (5) between the inlet and outlet points of the tubular receiver (ΔP) as:

$$\Delta P = (P_{in} - P_{out}) = \rho_f \times \xi \times \Delta x \frac{V_f^2}{2 \times D_{abs(int)}} \tag{6}$$

The quantity ξ, is the friction coefficient of the inner surface of the receiver tube.

The exergy lost owing to the heat transfer from absorber to the ambient can be expressed as in Ref. [9]:

$$\dot{E}_{ex_{loss,th}} = \sum_j \int_{T_{amb}}^{T_{abs,j}} \dot{Q}_{loss,j} \frac{T_{amb}}{T^2} dT = \sum_j \dot{Q}_{loss,j} (1 - T_{amb}/T_{abs,j}) \tag{7}$$

where, $\dot{Q}_{loss,j}$ are the total heat losses on the discretized element j.

Exergy destruction caused by irreversibilities in the solar collector systems is generally due to: (1) friction of viscous fluid and (2) identified heat transfer processes. Therefore, the exergy destruction can numerically be expressed as follows [9]:

$$\sum \dot{E}_{ex_{des}} = \sum \dot{E}_{ex_{des,\Delta P}} + \sum \dot{E}_{ex_{des,th}} \tag{8}$$

Exergy destructed due to pressure drop on the discretized element j (ΔP_j) during the heat transfer process occurring in the tubular receiver is expressed as in Ref. [34]:

$$\dot{E}_{ex_{des,\Delta P}} = T_{amb} \dot{m}_f \sum_j \frac{\Delta P_j \ln(T_{out,j}/T_{in,j})}{\rho_j (T_{out,j} - T_{in,j})} \tag{9}$$

It is well-known that exergy is also destroyed while heat is transferred from hot to cold medium. There are two identified heat transfer processes in the absorber that cause exergy rate destruction: (a) heat transfer of the solar energy absorbed by the surface of the absorber and (b) heat transfer conduction from outer absorber surface to fluid flow due to the temperature difference between the absorber surface and the HTF. Thus, we have the following relation [9]:

$$\sum \dot{E}_{ex_{des,th}} = \sum \dot{E}_{ex_{des,abs}} + \sum \dot{E}_{ex_{des,cond}} \tag{10}$$

The first term corresponds to the process occurring when the heat transfer takes place from the sun to the absorber surface, in this case, the entropy generation ($\dot{E}_{ex_{des,abs}}$) is defined as in [34]:

$$\dot{E}_{ex_{des,abs}} = \eta_{opt}\dot{E}_{ex_{in,r}}\left[1+\frac{1}{\psi}\left(\frac{\Delta x}{L}\sum_{j}\frac{T_{amb}}{T_{abs,j}}-1\right)\right] \tag{11}$$

In this equation, η_{opt} is the transient optical efficiency of the collector, ψ is the conversion efficiency of a radiation energy, it was derived independently by Petela in 1964 and later at the same year (1976) by Landsberg and Mallinson and Press; thus also call the PLP efficiency and given by [36, 37]:

$$\psi = 1-\frac{4}{3}\left(\frac{T_{amb}}{T_s}\right)+\frac{1}{3}\left(\frac{T_{amb}}{T_s}\right)^4 \tag{12}$$

As a parabolic trough solar collector can only concentrate direct normal radiation, PLP efficiency must take the directional character of this radiation into account. The effect of the sun's cone angle on the exergy content into the solar radiation was determined by Parrott and expressed as follows:

$$\psi = 1-\frac{4}{3}\left(\frac{T_{amb}}{T_s}\right)(1-\cos\delta_s)^{1/4}+\frac{1}{3}\left(\frac{T_{amb}}{T_s}\right)^4 \tag{13}$$

T_{amb} is the ambient temperature, $T_s{\approx}5800K$ is the equivalent temperature of the sun as a black body [37], and δ_s is the half angle of the cone subtended by the sun's disk ($\delta_s \sim 0.005$ rad) on a clear day [38]. The above PLP and Parrott expressions of the conversion efficiency apply just in the case of entirely concentrated solar radiation (concentration ratio about 46,200) and blackbody absorbers.

However, the experimental Sandia LS-2 collector considered in the present analysis has a small concentration ratio ($C = 22.42$), and his absorber is not blackbody but selective, characterized by an absorbance ($\alpha_{abs} = 0.96$) and an emittance ($\varepsilon_{abs} = 0.14$) as in Appendix B. Therefore other, more appropriate, exergy factors can be used as suggested by Badescu [38]. Then, the proposed correlation for the conversion efficiency of a radiation energy converter (operating between a high-temperature radiation reservoir of diluted radiation and a low-temperature heat reservoir) can be given as [38]:

$$\psi = \alpha_{abs}\left\{1-\frac{4}{3}\left(\frac{b}{l}\right)+\frac{1}{3}\left(\frac{b}{l}\right)^4\right\} \tag{14}$$

where $b = \frac{T_{amb}}{T_s}$ and l the interaction factor is defined by the following expression [38]: $l = \left\{\left(\frac{f_s}{f_a}\right)\times\left(\frac{\varepsilon_s}{\varepsilon_{abs}}\right)\times\alpha_{abs}\right\}^{1/4}$.

f_s is the flux density of the radiation energy incident on the absorber,

f_a is the geometric factor, ε_s is the dilution factor from the radiation reservoir, and ε_{abs} is the emittance of the absorber.

The quantity $\dot{E}_{ex_{in,r}}$ is defined as exergy related to the theoretical maximum useful work obtainable from solar energy with this device by the following equation [9, 34]:

$$\dot{E}_{ex_{in,r}} = a\times L\times I_d\times\psi \tag{15}$$

The second heat transfer process is between the absorber and the HTF. The expression of exergy destruction by heat transfer from the absorber to the fluid was proposed by Heugang et al. [9] as follows:

$$\dot{E}_{ex_{des,cond}} = T_{amb}\dot{m}_f A_f \left\{ \sum_j \left[\int_{T_{in,j}}^{T_{out,j}} C_{P,f}(T) \left(\frac{1}{T} - \frac{1}{T_{abs,j}} \right) dT \right] \right\} \quad (16)$$

The exergy analysis is concluded by the definition of the transient exergy efficiency, which is the ratio of gain exergy to solar irradiance exergy or the ratio of fluid exergy increment to exergy of solar irradiance normal to the collector aperture [9]:

$$\eta_{ex}(t) = \frac{\dot{E}_{ex_{gain,f}}}{\dot{E}_{ex_{in,r}}} \quad (17)$$

The overall daily average exergetic efficiency of the PTSC is expressed by [9, 33]:

$$\eta_{ex,d} = \int_{t_{sunrise}}^{t_{sunset}} \dot{E}_{ex_{gain,f}} dt \Big/ \int_{t_{sunrise}}^{t_{sunset}} \dot{E}_{ex_{in,r}} dt \quad (18)$$

2.4 Numerical methodology

The Gauss quadrature scheme is used to discretize the thermal and the exergy fluxes, which are deduced from energy and exergy analysis subsections and is also used to numerically evaluate integrals related to all the average quantities considered in the present study. The computational code is written in FORTRAN language compiler.

3. Results and discussion

The Sahelian region of Cameroon is the sample studied location (geographical coordinates: Latitude 5.15°, Longitude 13.58°, and Altitude 718 m). However, it is important to note while this research is using climatic conditions of Sahel, the value of this research extends beyond a Sahel African's zone itself as it enables the researchers to understand the comparative performances and behaviors of PTSCs under a real world and various climatic conditions. Therefore, the present numerical analysis by providing benchmark numerical results is a considerable theoretical support for further researches and engineering projects.

The Sahelian region of Cameroon is known as a Sudano-Sahelian tropical climate characterized by an irregular rainfall, a low cloudiness of the atmosphere, a hot and dry air, and the permanent aerosols, almost 10 sunniest months of dry seasons and two least sunny months of wet seasons. Then, two seasons are considered: 10 months of hot and dry season with the sunniest months and 2 months of wet season characterized by least sunny months with less sunny days.

3.1 The type of the heat transfer fluids

The HTFs considered in the present work are at temperatures that do not reach the phase change temperature, so that all the time, it is a question of sensible heat instead of latent heat. Any possibility of considering an eventual two-phase flow is excluded from this assumption. It is well-known that the liquid water has high density, high thermal conductivity, and low dynamic viscosity but its specific heat is higher compared to the same respective ones of the Syltherm 800 and the Therminol VP-1™ synthetic oils. Also, as the dry air is a perfect gas, the density, and the dynamic viscosity are almost null. The dry air has poor thermal conductivity, which means that the dry air behaves as a thermal insulator.

The problem treated in the present subsection is the effect of the HTF's type on the collector's behavior to determine the best fluids in the context of Sahelian climatic conditions. The simulation parameters are given as follows: the wind velocity around the collector is 2 m/s, the rate flow is $2.7 \times 10^{-4} kg/s.m^2$, the fluid inlet temperature is $20^{\circ}C$, and the tracking mode considered is the East-West polar. First of all, let considered **Figure 3**, which shows the incident direct solar irradiance in the sunny day of the sunniest month at the location where the study was carried. Also, **Figure 4** displays the transient temperatures of the ambient and sky at the two typical days considered. Let now consider **Figure 5**, displaying the transient outlet temperatures of HTFs, absorber pipe, and glass envelop for typical days of the Sahelian climatic conditions (16 and 21 March and 25 August). It is noted that during these two indicated typical days, the exit temperatures of the HTFs, absorber pipe, and glass envelop, all have the same pace in the shape of a bell. This observation is similar to the incident direct solar irradiance depicted in **Figure 4a** and the transient temperatures of ambient and sky at the considered typical days depicted in **Figure 5**. It shows the strong influence of the source term linked to direct solar irradiance in the transient thermal response of the glass envelop, the absorber pipe, and the HTFs according to their respective transient heat transfer equations. It also shows that these different subsystems (glass envelop, absorber pipe, and HTFs) are highly interdependent with their thermal behavior closely related. A comparison between the temperature profiles of different fluids used as HTFs is carried out. **Figures 5** and **6** show that synthetic oils have the highest temperatures, then follows dry air and finally liquid water; this classification can be observed at the outlet of the collector (temperature variation as a

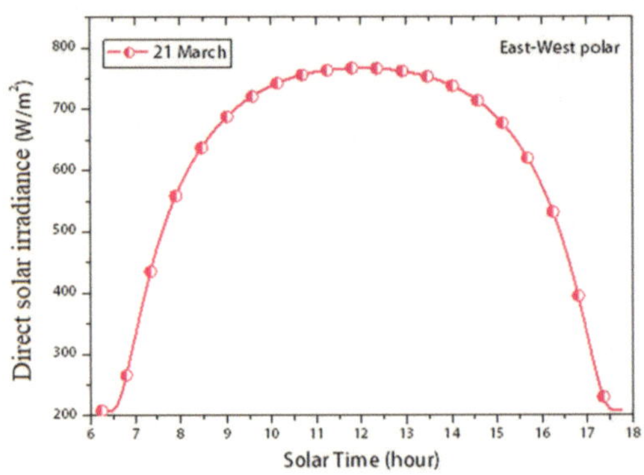

Figure 3.
Direct solar radiation on a typical sunny day of a sunniest month.

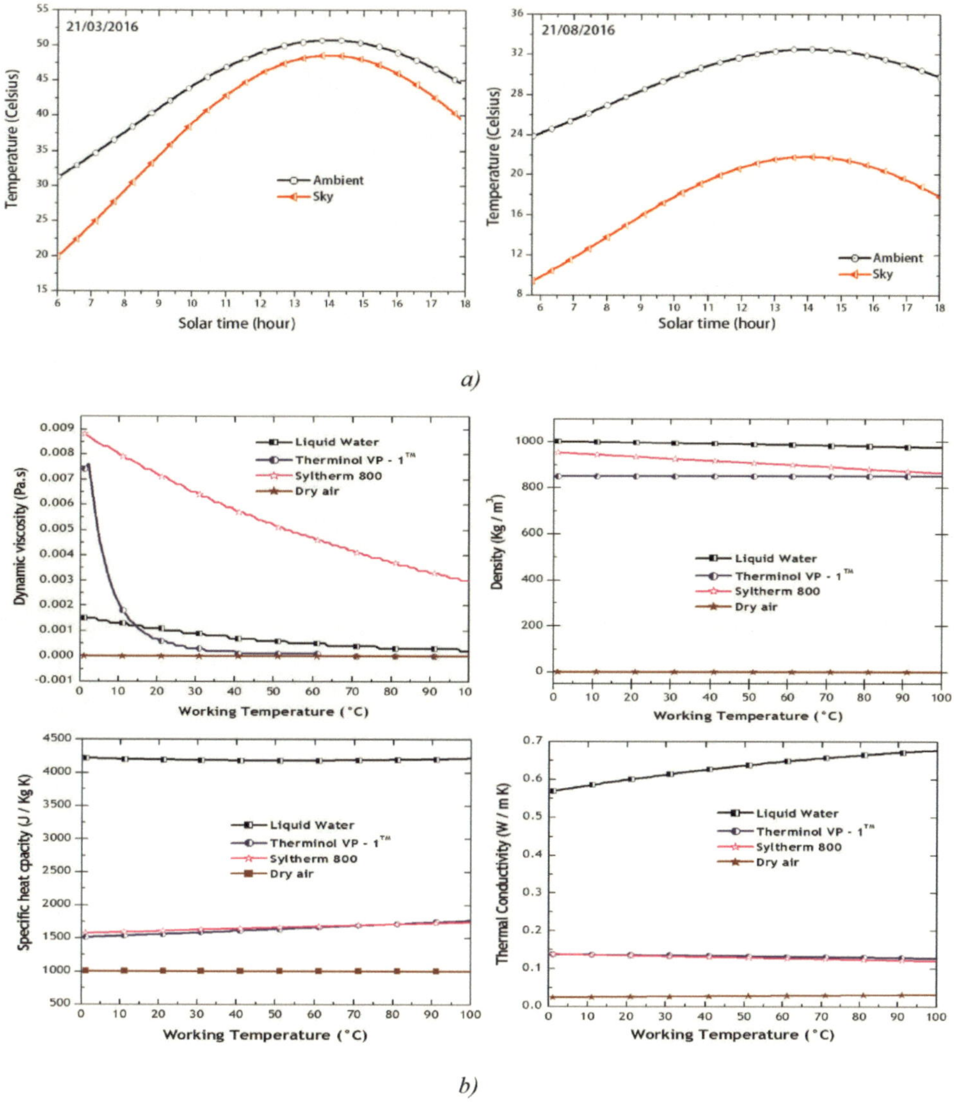

Figure 4.
(a) Transient temperatures of ambient and sky on two typical days of the Sahelian climatic conditions and (b) Thermophysical properties variation of HTFs with temperature.

function of time as presented in **Figure 5**) and along the tubular receiver (see **Figure 6**).

In an attempt to explain the temperature difference between the HTFs and the temperature differences between the glass envelop and the absorber pipe when the HTFs flow, two reasons seem plausible: the thermal stability of the physical phases of the HTFs and the effects of their thermophysical properties (see **Figure 4b**). The synthetic oils and water are in liquid phase; they have different phase change temperatures, which predispose them to operate in very different ways, even under the same conditions of thermal stress. The phase changing point of synthetic oils is about 400°C for Syltherm 800 and Therminol VP-1™, while that of water is 100°C. Thus, it is clear that the temperature of water in the physical liquid phase, without the appearance of

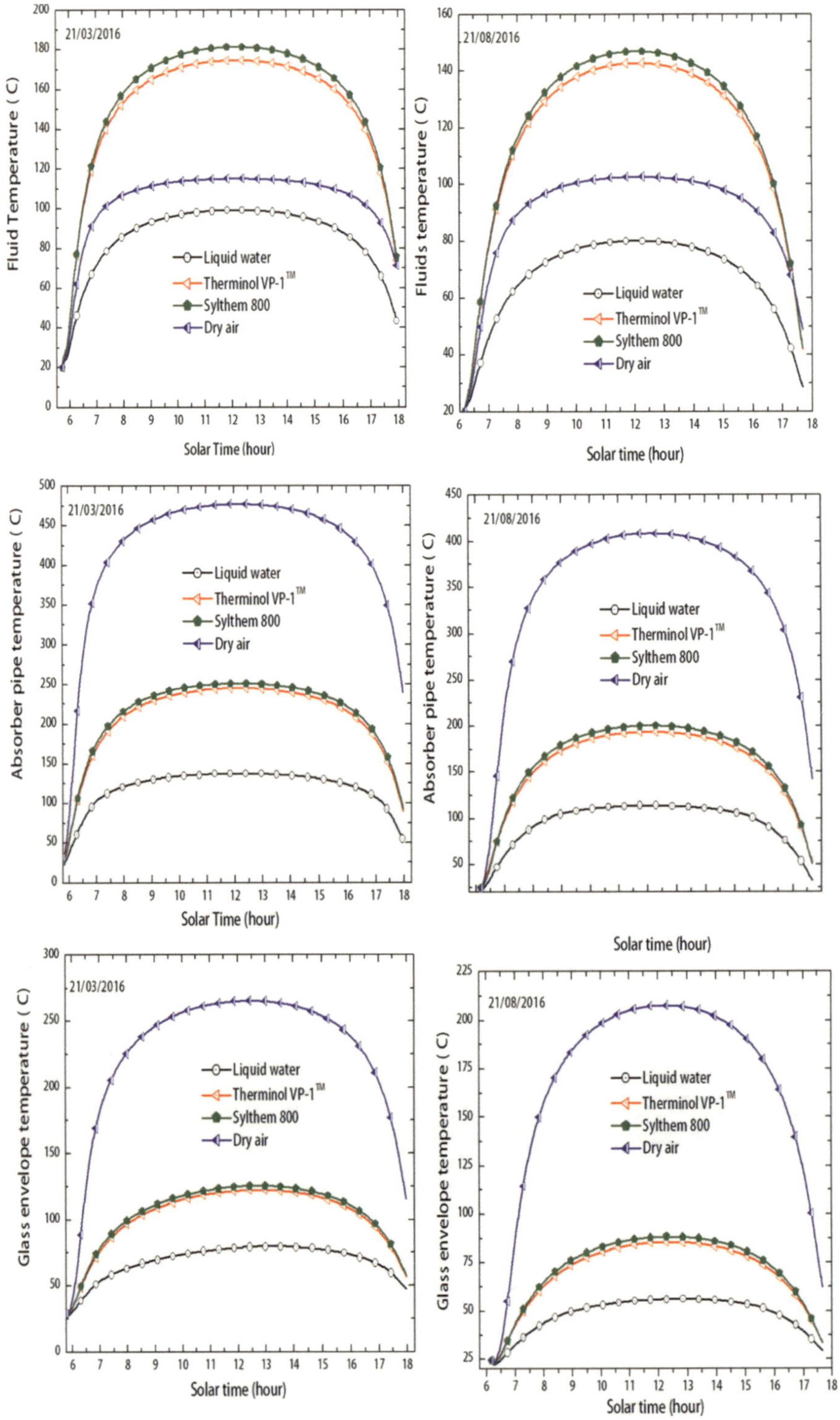

Figure 5.
Transient temperatures (at outlet of the collector): Fluid, absorber pipe, and glass envelop on two typical days of the Sahelian climatic conditions.

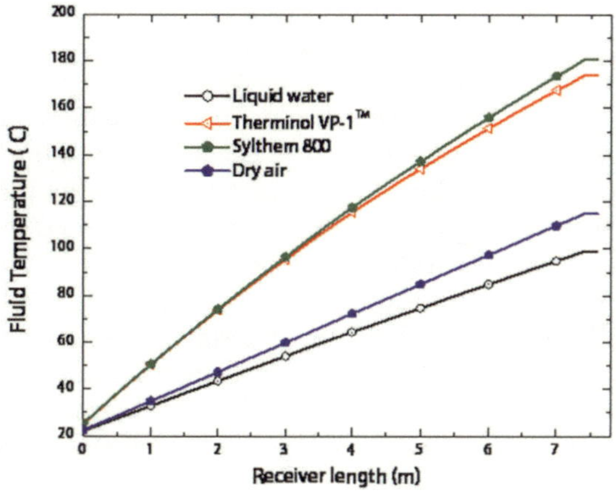

Figure 6.
Temperature distribution of the HTFs along the receiver tube at 13 h.

steam, cannot rise above 100°C. Therefore, its temperature profile is the lowest of all liquids on any given day.

On the other hand, the thermal agitation of the atoms and molecules of dry air, which increases under climatic conditions of the Sahel during the two typical days, explains why, when the entropy production increases the temperature of dry air is both above that of liquid water and below that of synthetic oils. **Table 1**, reporting the annual thermodynamic performances of PTSC using the different HTFs, summarizes all the quantitative details to better highlight the comparison. It is found that dry air produces the most significant internal heat loss and exergy loss by leakage heat transfer, 16,146.27 and 583,275.21 MWh per year, respectively, expressing too poor thermodynamic behavior of the PTSC, although the exergy destroyed by absorption or conduction is relatively moderate, compared to other HTFs. The argument on the thermophysical properties of fluids can be introduced to explain the higher or lower temperatures of the absorber pipe and of the glass envelop depending on the fluid, which flows in the absorber pipe. It is important to recall that this argument is based on the theory of the heat transfer into the pipe fluid flow. Indeed, the observation made is: the highest temperatures of the glass envelop and absorber pipe occur when

Annual values	Dry air	Liquid water	Therminol VP-1™	Syltherm 800
Exergy efficiency (%)	03.18	18.59	26.54	27.83
Thermal efficiency (%)	18.12	67.97	60.73	60.15
Internal geat loss (MWh)	16,146.27	1057.99	3311.62	3515.16
Exergy destructed by absorption (MWh)	443.29	6371.24	2960.36	2882.55
Exergy destructed by conduction (MWh)	4652.02	20,885.09	9068.09	10867.41
Exergy loss by heat leakage loss (MWh)	583,275.21	30,331.51	106,987.04	114246.28

Table 1.
Annual thermodynamic performances of PTSC with different working conditions using dry air, liquid water, Therminol VP-1™, and Syltherm 800 as heat transfer fluids.

the dry air circulates through the absorber pipe, secondly when it comes to synthetic oils, and finally when the liquid water is considered.

These high values of the temperatures of the glass envelop and absorber pipe are undesirable because they clearly express heat losses. While increasing the temperature of the glass envelop produces a high external heat loss, increasing the absorber tube temperature produces an internal heat loss. The temperatures of the glass envelope and the absorber pipe are all high as the fluid in question generates heat loss. The particular case of dry air can illustrate this. On sunny days, the maximum values of the temperatures of the glass envelop and the absorber pipe are of the order of 400 and 450°C, respectively. On less sunny days, these temperatures are about 200 and 260°C, respectively. It can be concluded that under the same thermal stress conditions, dry air being a gaseous fluid, and therefore atomically less thermally stable than liquids, heats up sufficiently for its temperature to exceed that of the phase change of water (boiling point). But the poor thermophysical properties of dry air, particularly its great inability to store thermal energy (illustrated by a poor specific heat capacity, accentuated with its increasing temperature as depicted by **Figure 4b**), lead to a considerably heat dissipation through the absorber pipe and the glass envelop. Thus, despite the thermal activities of its atoms and molecules, which are always more important, the temperature of dry air does not exceed those of synthetic oils (examples: absolute errors between the temperatures of dry air and those of Syltherm 800 and Therminol VP-1™ respectively, are for sunny day 62.5 and 58.93% and less sunny day 37.14 and 35.28%). Based on their more excellent thermal stability, synthetic oils

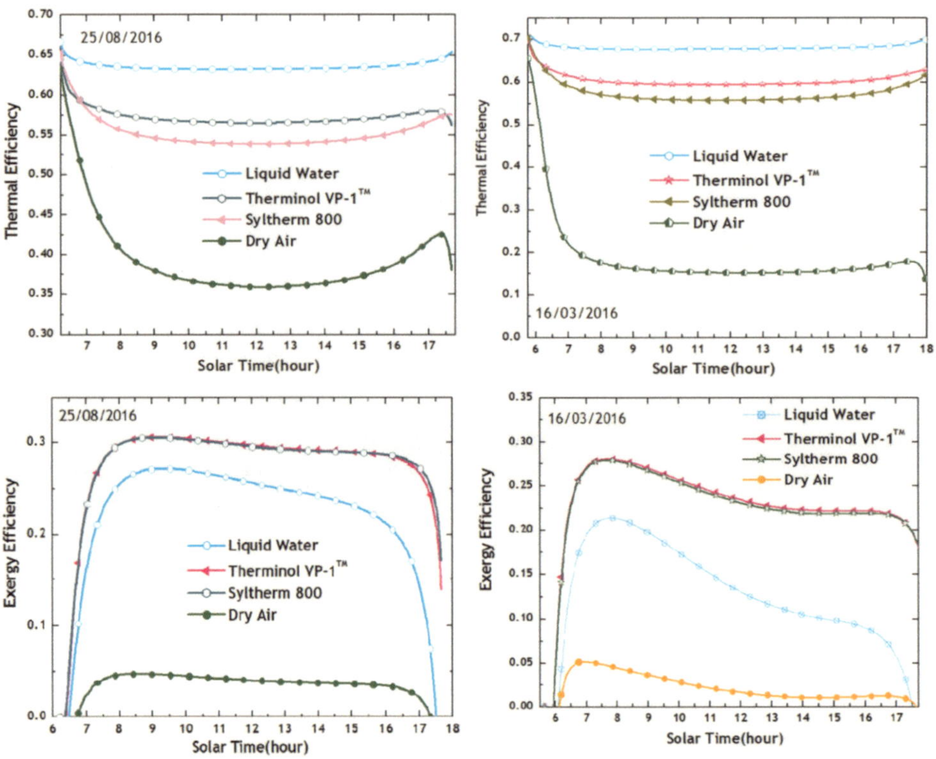

Figure 7.
Transient thermal and exergetic efficiencies of the collector for different HTFs on two typical days of the Sahelian climatic conditions.

can operate at temperatures around 450°C. To show the power of this conclusion, let an analysis based on efficiencies be introduced. For this reason, consider the unsteady thermal and exergy efficiencies at the two typical days (see **Figure** 7), and the monthly mean thermal and exergetic fluxes and efficiencies under conditions in Sahelian climate over an entire year (see **Figure** 7). The heat losses, with regard to **Figures** 7 and 8 and **Table 1**, are more accentuated in the case of dry air, which has the lowest thermal and exergetic efficiencies during the two typical days and even during all the year. The thermal agitation activities of the excited atoms and molecules of dry air produce a generation of entropy or a destruction of exergy in addition to thermal losses (as described above). In February, for example, the exergy is theoretically wholly destructed. The dry air does not produce any useful work in this month.

Regarding fluids in liquid phase, synthetic oils, and water, it is noted that water with better thermophysical properties achieves the best thermal efficiency during the two typical days and throughout the year with a monthly mean reaching up to 71.23% in September, and finally liquid water achieves the highest annual thermal efficiency, at about 67.97% (as shown in **Table 1**), under the numerical experimentation conditions specified at the beginning of this subsection. These results match with those established by Montes et al. [7] and Marif et al. [12], comparing thermal performances of PTSC in the presence of liquid water and Therminol VP-1™. The monthly mean exergy efficiency of water, which is about 26.52% in July, worse than those of synthetic oils, is explained more by a significant accumulation of destruction exergy flows. The first component is the exergy destruction due to the heat transfer from the absorber to the liquid water, caused by the temperature difference between the absorber surface and the agent fluid (with the highest annual value at about 20,885.09 MWh).

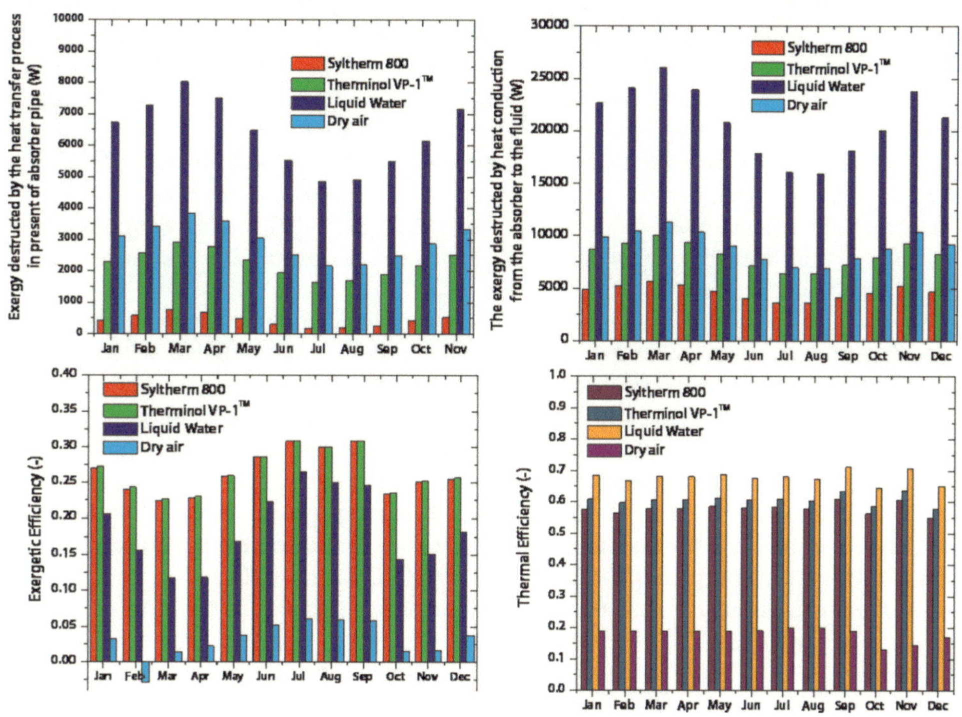

Figure 8.
Monthly mean daily thermal and exergetic (fluxes and efficiencies) for different HTFs in the Sahelian climatic conditions.

The other one is the exergy destruction due to the heat transfer of the solar energy absorbed by the surface of the absorber in the presence of liquid water (with also the highest annual value at about 6371.24 MWh). However, synthetic oils with their temperatures, the highest at the outlet of the collector, achieve the best exergy efficiency during the two typical days and throughout the year, with a maximum monthly mean value during the year equal to 32.3 and 30.87%, in September, respectively for Syltherm 800 and Therminol VP-1™.

Their low monthly mean thermal efficiency of about 63.66% for Syltherm 800 in November and 63.51% for Therminol VP-1™ in September, which are worse than water, is due to the heat losses and the exergy loss by heat leakage loss as reported in **Table 1**. These heat losses are external (at the level of glass envelop) and internal (at the level of absorber pipe). From these results, liquid water with its high thermal efficiency seems to be a good HTF. However, under the climatic conditions of the Sahel, its precarious thermal stability does not durably guarantee a single-phase flow necessary for the optimal functioning of the PTSC. The risk of the appearance of water vapor mixing with liquid water is almost permanent and can lead to the development of two-phases flows. Synthetic oils, on the other hand, which are much more thermally stable under harsh conditions of the Sahel, are more apt to work at very high temperatures and achieve attractive exergy efficiency. These results confirmed those published previously by Ouagued et al. [10] and Marif et al. [12] in the context of Sahara (Algeria), Kessel et al. [17] in the context of Cameroon Sahelian zones, and Zaharil et al. [25] in the Malaysian climatic condition under the steady-state regime. Ouagued et al. [10] compared various synthetic oils using as HTFs. They concluded that the best thermal oil is Syltherm 800; the Syltherm 800 can reach a temperature between 700 K and 800 K. Marif et al. [12] considered two fluids, liquid water, and ThermінolVP-1™ synthetic oil. In the case of synthetic oil, they found that high absorber pipe, glass envelop and output fluid temperatures compared with liquid water case. Nevertheless, the thermal efficiency of the PTSC with water was enhanced by 2% in all seasons to reach a maximum of 72.24% in summer. Kessel et al. [17] came to the conclusion that the maximum of outlet temperature of heat transfer fluid obtained at the right end of the absorber tube is about 140, 138, and 80°C during a typical day of the least sunny months and 180, 180, and 90°C during a typical day of the sunniest month respectively for water, air, and TherminolVP-1™ synthetic oil used as heat transfer fluid.

However, all these researchers have closely restricted their analyses to thermal modeling, and no information in relation to the thermodynamic behavior (or unsteady state exergetic modeling) of PTSC is provided.

On the contrary, Montes et al. [7] pointed out that direct steam generation using water is more efficient than oil and molten salt systems. They conclude that the annual performance seems to be also better for the case of direct steam generation technology as it has a slightly higher nominal overall efficiency than oil or salts. Zaharil et al. [25], proposing a steady-state energetic and exergetic analysis of Pressurized water, Therminol VP-1, Syltherm 800, Solar salt, Hitec XL, and liquid sodium under the Malaysian climate conditions found that liquid sodium outperformed all other HTFs considered at all temperature levels, except for the temperature level of 300 K with a maximum energetic and exergetic efficiency, respectively of 80 and 45.52%. They concluded that liquid sodium outperformed other HTFs between the range of 0.036–0.99% in energetic efficiency and 0.438–0.85% in exergetic efficiency.

3.2 The specific mass flow rate of the heat transfer fluid

The problem considered in this subsection is to examine how the specific mass flow rate of the HTF affects the thermodynamic behavior of the concentrator. The simulation parameters are: the wind velocity around the collector is 2 m/s, the inlet temperature of the HTF is 20°C, the tracking mode considered is the East-West polar, and the HTF considered is the Syltherm 800. The numerical experimentations were made at the two typical days of the Sahelian climatic conditions: the sunny day, 16 March and the less sunny day, 25 August. **Figure 8** demonstrates that the transient exergy efficiency decreases with the increase of the specific mass flow rate at the two typical days, but the values of the transient exergy efficiency are lower on the sunny day than its values on the less sunny day. It is also noticed that on the sunny day for $\dot{m}_f(kg/s.m^2) = 6.8 \times 10^{-4}$, the transient exergy efficiency runs to midday. After this time, everything else is canceled. So, it seems that for $\dot{m}_f(kg/s.m^2) > 6.8 \times 10^{-4}$, at the same date, the transient exergy efficiency will progressively but completely be canceled or will lead to a negative exergetic efficiency.

The meaning of this effect can be explained by the fact that exergy efficiency measures the exergy destruction and/or the exergy loss of the system due to irreversibilities. Yet, the second term of Eq. (5) of the exergy gained, which is negative, represents the decrease of mechanical energy due to flow friction (an irreversible process). This term increases in time with the specific mass flow rate and becomes

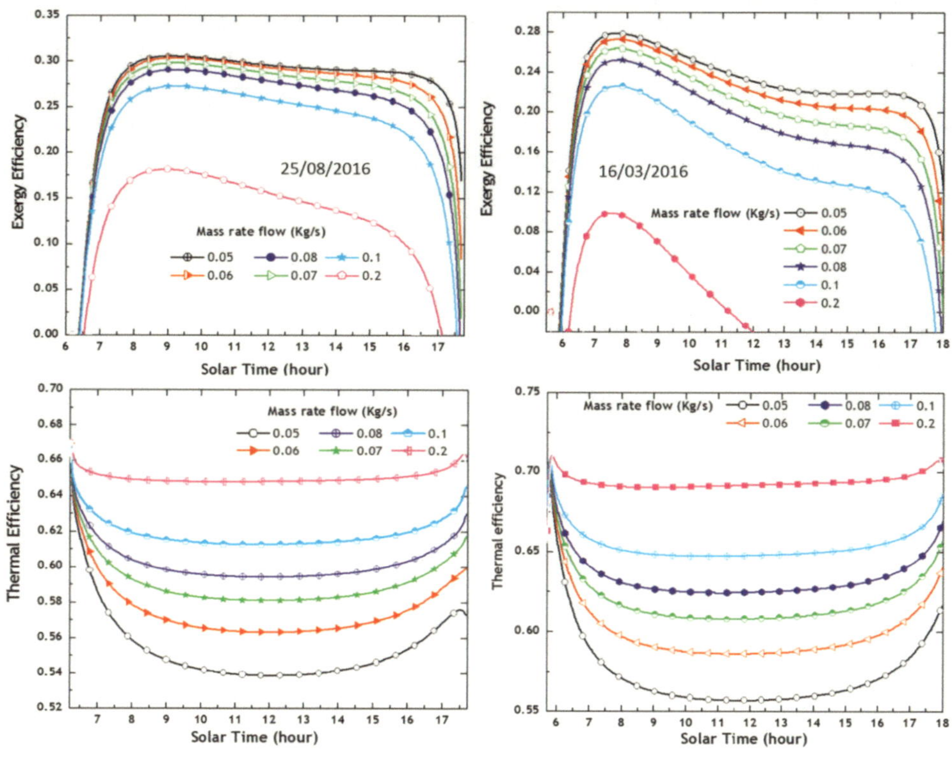

Figure 9.
Transient thermal and exergetic efficiencies of the collector for different mass flow rates on two typical days of the Sahelian climatic conditions.

considerably important, particularly on the sunny day, compared to the increase of other terms, with the increase of specific mass flow rate. To illustrate the dominant influence of mass flow rates on the exergy efficiency, it is considered the daily average based on the same implementation to obtain **Figure 8** of the exergy efficiency computed for some indicated values of the specific mass flow rate on these 2 days. For $\dot{m}_f(kg/s.m^2) = \{1.7, \ 2.7 \text{ and } 6.8\} \times 10^{-4}$, for example, on the sunny day the exergy efficiencies are respectively 25.13, 20.26, and -0.81% and on the less sunny day the exergy efficiencies are respectively 30.30, 27.57, and 14.43%. On the contrary, **Figure 9** shows that the transient thermal efficiency increases, while the specific mass flow rate increases these 2 days. In this case, the PTSC achieves its best performance on the sunny day with higher thermal efficiency than the one performed on the less sunny day. Indeed, it is established that the tube flow with a high specific mass flow rate induces moderated temperature of both absorber pipe and fluid, resulting in low heat losses. Let note the daily average of transient thermal efficiency, for the same range of the specific mass flow rate on these two typical days, $\dot{m}_f(kg/s.m^2) = \{1.7, \ 2.7 \text{ and } 6.8\} \times 10^{-4}$: on the sunny day the thermal efficiencies are respectively 59.31, 64.80, and 69.88% and on the less sunny day they are respectively 56.65, 61.13, and 65.31%. The effects of specific mass flow rates on the main thermodynamic performances parameters, such as thermal and exergy outputs (fluxes) and thermal and exergetic efficiencies throughout the year are reported in the **Table 2**. The decrease of the internal heat loss and exergy loss can be observed and the increase of the exergy destructed as a function of the specific masse flow rate too.

On one side, the internal heat loss and exergy loss vary respectively from 1,949.45 MW and 62,025.51 MWh for $2.7 \times 10^{-4} kg/s.m^2$ to 167.59 MWh and 3,624.77 MWh for $6.8 \times 10^{-3} kg/s.m^2$ by year. On the other side, the exergy destructed increases from 1,0019.98 MWh for $1.7 \times 10^{-4} kg/s.m^2$ to 54,566.68 MWh for $6.8 \times 10^{-3} kg/s.m^2$ by year. It can also be experienced that the observations made on the two typical days above remain the same concerning the opposite transient evolution of the thermal and exergetic efficiencies in relation to the specific mass flow rate. While the thermal efficiency increases from 32.60% for $1.7 \times 10^{-4} kg/s.m^2$ to 37.34% for $6.8 \times 10^{-3} kg/s.m^2$, the exergetic efficiency decreases from 12.89% for $1.7 \times 10^{-4} kg/s.m^2$ to a negative value -35.41% for $6.8 \times 10^{-3} kg/s.m^2$. At this stage, it is not superfluous to precise that for $1.71 \times 10^{-3} kg/s.m^2$ the annual average of thermal and exergetic efficiencies are respectively 37.7% and -16.9%. Thus, with the specific mass flow rates such as $1.71 \times 10^{-3} kg/s.m^2$ and $6.8 \times 10^{-3} kg/s.m^2$, verifying the condition

specific mass flow rate $\{10^{-5}\dot{m}_f(\text{kg/s.m}^2)\}$	Annual values				
	Fluxes (MWh)			Efficiency (%)	
	Internal Heat loss	Exergy loss	Exergy destructed	Thermal	Exergetic
17	1,949.45	62,025.51	10,019.98	32.60	12.89
27	1,104.61	33,085.99	13,956.52	35.08	09.59
68	834.47	24,104.99	16,122.14	35.86	07.32
17.1	113.25	2,539.32	36,368.29	37.70	—
680	167.59	3,624.77	54,566.68	37.34	—

Table 2.
Annual thermodynamic performances with different specific mass flow rates.

$\dot{m}_f(kg/s.m^2) > 6.8 \times 10^{-4}$, the explanations proposed in the second paragraph of this subsection in the case of exergy efficiency are valid here in the case of exergy destructed; it is convenient to underline the similitude of the respective expressions. Moreover, the other irreversible processes in the system responsible of exergy destruction are (a) the heat transfer from the absorber pipe to the HTF at finite temperature difference noted $\dot{E}_{ex_{des,th}}$; (b) the viscous friction fluid between the collector fluid and the pipe due to pressure drop during this process (Eq. (9)). This last flux is often null in situations involving very low mass flow rate ($\dot{m}_f(kg/s.m^2) \sim 2.7 \times 10^{-4}$). But for high value of specific mass flow rate, this flux becomes considerable. Computations have shown that the annual values of exergy destructed due to friction between the collector fluid and the pipe is important, especially 4.9×10^{-3} MWh for $1.71 \times 10^{-3} kg/s.m^2$ and 35.87×10^{-2}MWh for $6.8 \times 10^{-2} kg/s.m^2$ under the conditions used in the above numerical simulations. These results are in accordance with those outlined previously by two other research groups: Padilla et al. [34] and Kumar and Kumar [39]. The first, that is to say, Padilla et al. [34] investigated the steady-state energy and exergy performance analysis using Syltherm 800 of the SEGS LS-2 parabolic trough solar collector. They concluded that the optimal performance of PTSC is independent of the specific mass flow rate, since the exergy destruction is due to: (1) friction of the heat transfer fluid, and (2) heat transfer between the absorber and the HTF. However, their contribution to the total exergy wasted is less than 0.5%. The group from Kumar and Kumar [39] investigated the steady-state thermal behavior of a PTSC using rather water. They found that collectors' performance chiefly depends upon the specific mass flow rate, and no considerable change is found when the specific mass flow rate of water is more than $8.21 \times 10^{-5} kg/s.m^2$. In the present study, Syltherm 800 is considered; the limit of the specific mass flow rate reached in our case is $6.8 \times 10^{-4} kg/s.m^2$, below which the exergy efficiency cancels and thermal efficiency changes very slowly. Also, Marefati et al. [40] concluded (as in the present study) their thermal analysis of the PTSC for four cities of Iran with different weather conditions (Shiraz, Yazd, Tehran and Tabriz) in different months, by stating that efficiency and output useful energy of the collector increase by specific mass flow rate from $3{\times}10^{-2} kg/s.m^2$ to $5{\times}10^{-2} kg/s.m^2$. However, the collector outlet temperature decreases by specific mass flow rate. By contrast, by decreasing the specific mass flow rate from $3{\times}10^{-2}\ kg/s.m^2$ to $1{\times}10^{-2} kg/s.m^2$, the collector's efficiency, and useful energy output decrease, but collector output temperature increases.

3.3 Wind velocity effect on the heat transfer around a tubular receiver

This section investigates the problem of convection heat losses to the ambient as an effect of wind flow on the heat transfer around the PTSC structure. So, the results of numerical simulations of wind velocity effects on the heat transfer close to the HCE are presented and discussed. Since the absorber is enclosed in a glass envelop, the glass cover insulates the absorber surface from convective heat loss due to wind. Therefore, effectiveness of the glass cover can be tested by varying the wind velocity and computing specific parameters of the PTSC for changes. The transient key parameters displayed in this study are the external heat losses, the exergy loss due to heat transfer from absorber to the ambient, the heat loss coefficient, and the thermal and exergy efficiencies. This problem has received the great attention of researchers, such as

Hachicha [28]. The simulation parameters are: the specific mass flow rate of $3 \times 10^{-3} kg/s.m^2$, the inlet temperature of the HTF at 20°C, the tracking mode considered at the East-West polar, and the HTF considered is the Syltherm 800. A uniform wind velocity in the inlet velocity profile is assumed, and the different used values of the wind velocity are $V_{wind}(m/s) = \{0, 1, 4, 7 \text{ and } 12\}$. The numerical simulations are made on the two typical days of the Sahelian climatic conditions: the sunny day, 16 March and the less sunny day, 25 August; then depending on these days and meteorological conditions, convection around the HCE might be free or forced. **Figure 10** shows the transient variation of the heat transfer from a PTSC in crossflow to analyze the influence of wind velocity. It can be noticed that the transient external heat loss increases with the wind velocity on the two typical days. However, the transient thermal and exergetic efficiencies, which have the same time variation, decrease with

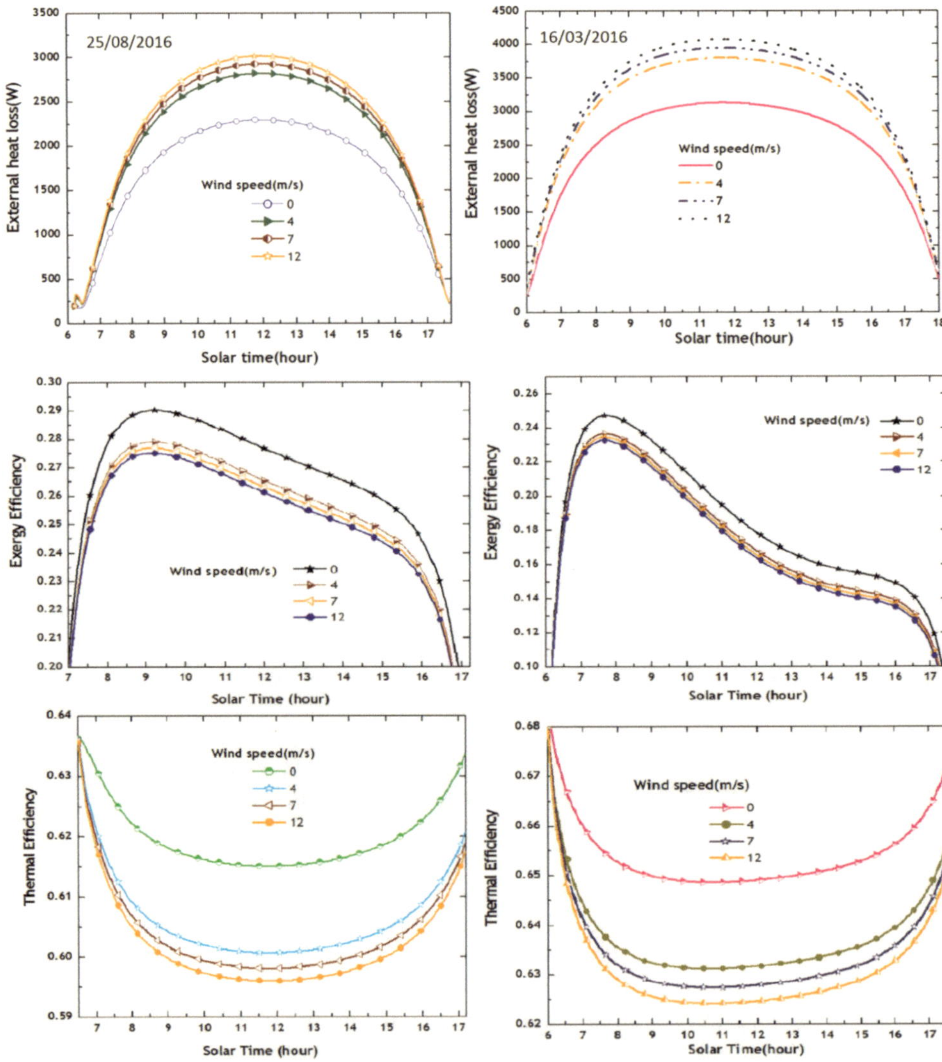

Figure 10.
Transient external heat loss and exergetic and thermal efficiencies of the collector for different wind velocities on two typical days of the Sahelian climatic conditions.

wind velocity (m/s)	Annual values			Overall daily average values of Efficiencies (%)			
	External Heat loss (MWh)	Thermal Efficiency (%)	Exergy Efficiency (%)	Thermal		Exergy	
				Sunny day	Less sunny day	Sunny day	Less sunny day
0	3,004.07	71.24	25.17	66.81	62.85	18.99	27.07
1	3,304.60	71.19	24.60	66.20	62.28	18.61	26.66
4	3,710.31	71.12	24.18	65.40	61.63	18.08	26.13
7	3,856.33	71.10	23.97	65.11	61.42	17.88	25.94
12	3,977.56	71.08	23.80	64.88	61.24	17.71	25.78

Table 3.
Wind velocity effects on the heat transfer around a PTSC in Sahelian climatic conditions.

the increase of the wind velocity. Consider now the overall daily values of these efficiencies the 16 March and the 25 August, the two named typical days, reported in **Table 3**. It can be seen that for $V_{wind}(m/s) = \{0, 1, 4, 7 \text{ and } 12\}$, the thermal efficiency values are respectively 66.81, 66.20, 65.40, 65.11, and 64.88% on the sunny day and 62.85, 62.28, 61.63, 61.42, and 61.24% on the less sunny day; the exergy efficiency values are respectively 18.99, 18.61, 18.08, 17.88, and 17.71% on the sunny day and 27.07, 26.66, 26.13, 25.94, and 25.78% on the less sunny day. **Table 3** also shows that the variations remain the same by considering the annual average values of the parameters to the variation of wind velocity.

Let illustrate these observations by presenting the increase of the annual values of the external heat loss in the same range of the wind velocity. The annual values of the external heat loss are: 3,004.07, 3,304.60, 3,710.31, 3,856.33, and 3,977.56 MWh by year, respectively for $V_{wind}(m/s) = \{0, 1, 4, 7 \text{ and } 12\}$; thus, it can be concluded that the variation described above, that is. external heat loss vs. wind velocity, leads to an increment of 24.47%. In the cases of thermal and exergetic efficiencies, the decrement, following the same rhythm for the given wind velocity: $V_{wind}(m/s) = \{0, 1, 4, 7 \text{ and } 12\}$, are respectively 0.22 and 5.44%. It can be concluded that the annual values of the efficiencies do not vary significantly in the range of the wind velocity considered.

For **Figure 9**, the specific mass flow rate considered is $1.03 \times 10^{-1}\ kg/s.m^2$ on the least sunny month. The considered values of the wind velocity for this numerical experimentations, in these cases, are $V_{wind}(m/s) = \{0, 0.5, 1, 2, 4, 7, 10 \text{ and } 12\}$. It can be seen from **Figure 10** that the transient thermal efficiency globally decreases with the increase of the ambient wind speed, while the transient heat loss coefficient increases with ambient wind speed. However, beyond $V_{wind}(m/s) \geq 2$, the variation of the exergy loss is slow and tends to be insignificant with the increase of the ambient wind velocity. This observation confirms the other one made before, concerning the slow evolution of the overall daily and annual average values of the thermal and exergetic efficiencies reported in **Table 3**. This result obtained in the case of vacuum in the annulus receiver tube is in agreement with those published by Padilla et al. [34] and Zaharil et al. [25]. Padilla et al. [34] found that vacuum in annulus mitigates the effect of wind speed. However, its absence increases the thermal losses to the surroundings, which reduces both thermal and exergy efficiencies, and they concluded that the effect of wind speed is negligible. Zaharil et al. [25] concluded that wind speed increment has a negligible impact on PTSC's performance for all HTFs, with liquid sodium being the most resistant to any wind increment with reduction of only 0.0482 and 0.0482276% for energetic and exergetic efficiency, respectively, when wind velocity increases from 2 m/s to 18 m/s.

4. Conclusion

In the present study, the effective approach recently developed in [9], based on transient both thermal and exergetic performances of the Sandia National Laboratory, is applied with the view to examine the sensitivity of the thermal behavior of a PTSC to the following parameters in the Sahelian climatic conditions: the collector parameters, including the type of the HTF, the operational parameters, related to specific mass flow rate of the HTF, and the wind velocity around the collector. The energetic and exergetic analysis provided the following results:

- Concerning the working fluids selection, the Syltherm 800, which is more thermally stable, among Therminol VP-1™, liquid water, and dry air, is more apt to work at very high temperature at about $450^{\circ}C$ and achieves overall annual daily average of thermal and exergy efficiencies, respectively 60.15 and 27.83%. However, it must be taken into account that these values have been calculated according to very specific assumptions, such as the solar-only plant without power cycle and thermal storage.

- Concerning the effects of specific mass flow rate, the exergy destructed increases from 10,019.98 MWh for $1.7 \times 10^{-4} kg/s.m^2$ to 54,566.68 MWh for $6.8 \times 10^{-3} kg/s.m^2$ by year; the internal heat loss and the exergy loss vary respectively from 1,949.45 and 62,025.51 MWh for $1.7 \times 10^{-4} kg/s.m^2$ to 167.59 and 3,624.77 MWh for $6.8 \times 10^{-3} kg/s.m^2$ per year. Exergy and thermal efficiency have an opposite transient evolution. In addition, for specific mass flow rate $\dot{m}(kg/s.m^2) > 6.8 \times 10^{-4}$, the exergy efficiency decreases progressively, and completely cancels or leads to negative value. However, these values have been calculated according to a specific assumption: the solar-only plant without power cycle and thermal storage, is considered. Thus, other engineering issues such as control, materials, and the economic analysis must account in the HTF technology evaluation.

- About the impacts of the wind velocity on the collector heat transfer, the annual values of the external heat loss are: 3,004.07, 3,304.60, 3,710.31, 3,856.33, and 3,977.56 MWh per year, respectively for $V_{wind}(m/s) = \{0, 1, 4, 7 \text{ and } 12\}$; thus, the variation described above i.e., external heat loss vs. wind velocity, leads to an increment of 24.47%. In the cases of thermal and exergetic efficiencies, and following the same rhythm for the given wind velocity: $V_{wind}(m/s) = \{0, 1, 4, 7 \text{ and } 12\}$, the decrement is respectively 0.22 and 5.44%. It can be concluded that the annual values of the efficiencies do not vary significantly in the range of the wind velocity considered.

Acknowledgements

The authors thank the editor and reviewers for useful comments and suggestions. The authors would like to convey thanks to R. Tchitnga, Department of Physics, University of Dschang, Cameroon, for his valuable proofreading of the manuscript.

Conflict of interest

The authors declare no conflict of interest.

Abbreviations

CFD	Computational Fluid Dynamics
CSP	Concentrating solar power
EES	Engineering equation solver
HCE	Heat collector element
HTF	Heat transfer fluid
LBE	Lead-Bismuth Eutectic
MWh	Megawatt hour
PLP	Petela Landsberg Press
PTSC	Parabolic trough solar collector
SEGS	Solar Energy Generation System
vs.	Versus

Nomenclature

a	Aperture,	m
A	Area,	m^2
C	Solar concentration ratio,	$(-)$
C_{p_f}	Specific heat at constant pressure of fluid,	$kJ/(kg\,K)$
$\dot{E}$	Exergy flux,	W
I_d	Direct solar irradiance,	W/m^2
j	Surface, Element,	$(-)$
L	Length of the parabolic trough solar collector,	m
$\dot{m}$	Specific mass flow rate,	$kg/s.m^2$
$\dot{Q}$	Heat flux,	W
T	Temperature,	$^{\circ}C$
V_f	Fluid velocity,	m/s

Greek symbols

Δx	One-dimensional control volume,	m
η	Efficiency,	(%)
μ	Dynamic viscosity,	Pa•s or kg/m•s

Subscripts

abs	Absorber
Amb	Ambient
Carnot	Carnot
Cond	Conduction
D	Direct; the overall daily value of optical, thermal, and exergetic efficiency
Dest	Destructed
Ex	Exergy
F	Fluid or Heat transfer fluid (HTF)
G	Glass
In	Inlet, input
L	Loss
Opt	Optical
Out	Outlet, output
P	Pressure
R	Irradiance
Th	Thermal
wind	Wind

Superscripts

T	Previous time step
$t+\Delta t$	Present time step

Appendixes

Appendix A

See **Table 4**.

Water [12]
$\rho_{water} = -495{,}626 \times 10^{-9}T^2 - 23{,}291 \times 10^{-5}T + 1{,}001.83T$
$K_{water} = -596{,}341 \times 10^{-11}T^2 + 1.68 \times 10^{-3}T + 56{,}821 \times 10^{-5}$
$Cp_{water} = 1{,}378 \times 10^{-4}T^2 - 142{,}026 \times 10^{-5}T + 4128.23$
$\mu_{water} = -4{,}282.65 \times 10^{-13}T^3 + 1{,}889.79 \times 10^{-10}T^2 - 277.74 \times 10^{-7}T + 15.6 \times 10^{-4}$
Dry air [12]
$\rho_{air} = 353/(T + 237.15)$
$K_{air} = 7.57 \times 10^{-5}T^2 + 242 \times 10^{-4}$
$Cp_{air} = 1{,}008$
$\mu_{air} = 10^{-9}(46T + 17{,}176)$
Therminol VP-1™ [12]
$\rho_{oil} = 850$

$K_{oil} = -1{,}729 \times 10^{-10}T^2 - 870.8 \times 10^{-7}T + 1{,}381 \times 10^{-4}$
$Cp_{oil} = 788.8 \times 10^{-6}T^2 + 2{,}496 \times 10^{-3}T + 1{,}509$
$\mu_{oil} = 10^3\left(10^{0.8703} \times T^{\left(0.2877+Log\left(T^{-0.3618}\right)\right)}\right)$
Syltherm 800 [9, 41]
$\rho_{oil} = 12{,}690{,}306 \times 10^{10} - 152{,}080{,}898 \times 10^{-8}(T+273) + 179{,}056{,}367 \times 10^{-11}(T+273)^2 - 167{,}068{,}252 \times 10^{-14}(T+273)^3$
$K_{oil} = 19{,}011{,}994 \times 10^{-8} - 188{,}022{,}387 \times 10^{-12}(T+273)$
$Cp_{oil} = 11{,}078{,}577 \times 10^{10} + 170{,}742{,}274 \times 10^{-8}(T+273)$
$\mu_{oil} = 8{,}486{,}612 \times 10^{-8} - 5{,}541{,}277 \times 10^{-10}(T+273) + 1{,}388{,}285 \times 10^{-12}(T+273)^2 - 1{,}566{,}003 \times 10^{-15}(T+273)^3 + 6{,}672{,}331 \times 10^{-18}(T+273)^4$

Table 4.
Temperature (°C)-dependent thermophysical properties of used heat transfer fluids.

Appendix B

See **Table 5.**

Designation of the parameter (symbol)	Value, unity
Receiver length (L)	$7.8m$
Collector width or aperture (a)	$5m$
Focal distance (F)	$1.84m$
Reflector surface reflectivity Shape (ρ_0)	0.93×10^{-1}
Shape factor (f)	0.92×10^{-1}
Solar concentration ratio (C)	22.42
Absorber pipe material	*Cermet selective surface*
Absorber pipe external diameter $(D_{abs(ext)})$	$7 \times 10^{-2}m$
Absorber pipe internal diameter $(D_{abs(int)})$	$6.6 \times 10^{-2}m$
Absorber pipe thermal conductivity (K_{abs})	$54W/mK$
Absorbed pipe emittance (ε_{abs})	1.4×10^{-1}
Absorber pipe thermal absorbance (α_{abs})	9.6×10^{-1}
Absorber pipe specific heat (C_{pabs})	$500J/kg.K$
Absorber pipe density (ρ_{abs})	$8{,}020kg/m^3$
Transmittance-absorbance (α_0)	8.64×10^{-1}
Glass envelop material	*Pyrex*
Glass envelop external diameter $(D_{g(ext)})$	$1.15 \times 10^{-1}m$
Glass envelop internal diameter $(D_{g(int)})$	$1.09 \times 10^{-1}m$
Glass envelop thermal conductivity (K_g)	$1.2W/mK$
Glass envelop emittance (ε_g)	8.6×10^{-1}

Designation of the parameter (symbol)	Value, unity
Glass envelop thermal absorbance (α_g)	2×10^{-2}
Glass envelop specific heat (C_{Pg})	$1{,}090 J/kg.K$
Glass envelop density (ρ_g)	$2{,}230 kg/m^3$
Glass envelop transmittance (τ_g)	9.5×10^{-1}
Receiver	*Evacuated tube, metal bellows at each end*

Table 5.
Geometrical, optical, and thermophysical data for the LS-2 parabolic trough solar collector [9].

Author details

Steven Audrey Heugang Ndjanda[1] and Etienne Tchoffo Houdji[2,3]*

1 Laboratory of Solar Energy, African Institute of Sustainable Energy Technology, Expertise for Africa, Yaoundé, Cameroon

2 Department of Renewable Energy, Advanced School of Mines Processing and Energy Resources, The University of Bertoua, Cameroon

3 Department of Renewable Energy, National Advanced School of Engineering, The University of Maroua, Cameroon

*Address all correspondence to: tchoffohoudji@yahoo.fr

References

[1] Kalogirou AS. Solar thermal collectors and applications. Progress in Energy and Combustion Science. 2004;**30**:231-295. DOI: 10.1016/j.pecs.2004.02.001

[2] Nixon JD, Dey PK, Davies PA. Which is the best solar thermal collection technology for electricity generation in north-West India? Evaluation of options using the analytical hierarchy process. Energy. 2010;**35**:5230-5240. DOI: 10.1016/j.energy.2010.07.042

[3] Ehtiwesh ASI, Neto DSF, Sousa CMA. Deployment of parabolic trough concentrated solar power plants in North Africa—A case study for Libya. International Journal of Green Energy. 2019;**16**(1):72-85. DOI: 10.1080/15435075.2018.1533474

[4] Sueyoshi T, Goto M. Comparison among three groups of solar thermal power stations by data envelopment analysis. Energies. 2019;**12**(2454):1-20. DOI: 10.3390/en12132454

[5] Yudong L, Fangqin L, Jianxing R, Guizhou R, Honghong S, Gang L. Solar thermal power generation technology research. E3S Web of Conferences. 2019; **136**:02016. DOI: 10.1051/e3sconf/201913602016

[6] Taylor N. Solar Thermal Electricity: Technology Development Report. Luxembourg: EUR 29913 EN, Publications Office of the European Union; 2019. DOI: 10.2760/67261 ISBN 978-92-76-12535-8. JRC118040

[7] Montes JM, Abandas A, Martinez-Val MJ. Thermofluidynamic model and comparative analysis of parabolic trough collectors using oil, water/steam, or molten salt as heat transfer fluids. Journal of Solar Energy Engineering. 2010;**132**:1-7. DOI: /10.1115/1.4001399

[8] Bhargav HU, Amitkumar JP, Ramana PV. A detailed review on solar parabolic trough collector. International Journal of Ambient Energy. 2019;**43**:176-196. DOI: 10.1080/01430750.2019.1636869

[9] Heugang NAS, Kamdem THT, Tchoffo HE, Pelap FB. Transient energy and exergy analysis of parabolic trough solar collector with an application to Sahel climate. International Journal of Sustainable Energy. 2020;**40**(6):557-583. DOI: 10.1080/14786451.2020.1828418

[10] Ouagued M, Abdallah K, Larbi L. Estimation of the temperature, heat gain and heat loss by solar parabolic trough collector under Algerian climate using different thermal oils. Energy Conversion and Management. 2013;**75**:191-201. DOI: 10.1016/j.enconman.2013.06.011

[11] Sa M, Zaversky F, Medina R, Garcı J, Astrain D. Object-oriented modelling for the transient performance simulation of parabolic trough collectors using molten salt as heat transfer fluid. Solar Energy. 2013;**95**:192-215. DOI: 10.1016/j.solener.2013.05.015

[12] Marif Y, Hocine B, Hamza B, Mohamed MB, Moussa Z. Numerical simulation of solar parabolic trough collector performance in the Algeria Saharan region. Energy Conversion and Management. 2014;**85**:521-529. DOI: 10.1016/j.enconman.2014.06.002

[13] Hussein AK. Applications of nanotechnology to improve the performance of solar collectors – Recent advances and overview. Renewable and Sustainable Energy Reviews. 2016;**62**:767-792. DOI: 10.1016/j.rser.2016.04.050

[14] Ghoneim AA, Mohammedein AM. Parabolic trough collector performance in a hot climate. Journal of Energy Engineering. 2016;**142**(1):1-11. DOI: 10.1061/(ASCE)EY.1943-7897.000026

[15] Najjar YSH, Sadeq J. Modeling and simulation of solar thermal power system using parabolic trough collector. Journal of Energy Engineering. 2017; **143**(2):1-9. DOI: 10.1061/(ASCE)EY.1943-7897.0000404

[16] Wan Y, Zhu Y, Chen H, Yang L, Yang M. Thermal performance of a single-pass all-glass parabolic trough receiver. Journal of Energy Engineering. 2016;**143**(1):1-7. DOI: 10.1061/(asce)ey.1943-7897.0000381

[17] Kessel JP, Haman-Djalo BT, Tchoffo EH, Heugang SN. Optical and thermal performances of a solar parabolic trough collector under climate conditions of the Cameroon Sahelian regions. International Journal of Innovative Science and Research. 2017; **29**(2):149-165

[18] Mwesigye A, Yılmaz İH. Thermal and thermodynamic benchmarking of liquid heat transfer fluids in a high concentration ratio parabolic trough solar collector system. Journal of Molecular Liquids. 2020;**319**(1):114151. DOI: 10.1016/j.molliq.2020.114151

[19] Okonkwo EC, Essien EA, Abid M, Kavaz D, Ratlamwala TA. Thermal performance analysis of a parabolic trough collector using water-based green-synthesized nanofluids. Solar Energy. 2018;**170**:658-670. DOI: 10.1016/j.solener.2018.06.012

[20] Bellos E, Tzivanidis C. Thermal efficiency enhancement of nanofluid based parabolic trough solar collectors. Journal of Thermal Analysis and Calorimetry. 2019;**135**:597-608. DOI: 10.1007/s10973-018-7056-7

[21] Evangelisti L, De Lieto RV, Asdrubali F. Latest advances on solar thermal collectors: A comprehensive review. Renewable and Sustainable Energy Reviews. 2019;**114**:1-20. DOI: 10.1016/j.rser.2019.109318

[22] Okonkwo EC, Abid M, Ratlamwala TAH. Comparative study of heat transfer enhancement in parabolic trough collector based on modified absorber geometry. Journal of Energy Engineering. 2019;**145**(3):1-16. DOI: 10.1061/(ASCE)EY.1943-7897.0000602

[23] Al-Oran O, Lezsovits F, Aljawabrah A. Exergy and energy amelioration for parabolic trough collector using mono and hybrid nanofluids. Journal of Thermal Analysis and Calorimetry. 2020;**140**:1579-1596. DOI: 10.1007/s10973-020-09371-x

[24] Abed N, Afgan I, Cioncolini A, Iacovides H, Nasser A. Assessment and evaluation of the thermal performance of various working fluids in parabolic trough collectors of solar thermal power plants under non-uniform heat flux distribution conditions. Energies. 2020; **13**:1-27. DOI: 10.3390/en13153776

[25] Zaharil HA, Hasanuzzaman M. Modelling and performance analysis of parabolic trough solar concentrator for different heat transfer fluids under Malaysian condition. Renewable Energy. 2020;**149**:22-41. DOI: 10.1016/j.renene.2019.12.032

[26] Yılmaz HI, Mwesigye A. Modeling, simulation and performance analysis of parabolic trough solar collectors: A comprehensive review. Applied Energy. 2018;**225**:135-174. DOI: 10.1016/j.apenergy.2018.05.014

[27] El Mghouchi Y, El Bouardi A, Choulli Z, Ajzoul T. Models for obtaining the daily direct, diffuse and global solar radiations. Renewable and Sustainable Energy Reviews. 2016;**56**:87-99. DOI: 10.1016/j.rser.2015.11.044

[28] Hachicha AA. Numerical Modelling of a Parabolic Trough Solar Collector. Doctorate thesis in physics,. Spain: Universitat Politècnica de Catalunya; 2013. pp. 1-177. DOI: 10.5821/dissertation-2117-95129

[29] Paetzold J, Cochard S, Vassallo A, Fletcher DF. Wind engineering analysis of parabolic trough solar collectors: The effects of varying the trough depth. Journal of Wind Engineering and Industrial Aerodynamics. 2014;**135**: 118-128. DOI: 10.1016/j.jweia.2014.10.017

[30] Fu W, Yang MC, Zhu YZ, Yang L. The wind-structure interaction analysis and optimization of parabolic trough collector. Energy Procedia. 2015;**69**: 77-83. DOI: 10.1016/j.egypro.2015.03.010

[31] Hassane MB, Njomo D, Barka M, Youssouf MK, Goron D, Chara-Dackou VS, et al. Modeling the incident solar radiation of the City of N'Djamena (Chad) by the Capderou method. International Journal of Photoenergy. 2020;**6292147**:1-10. DOI: 10.1155/2020/6292147

[32] Eskin N. Transient performance analysis of cylindrical parabolic concentrating collectors and comparison with experimental results. Energy Conversion and Management. 1999;**40**: 175-191. DOI: 10.1016/S0196-8904(98)00035-1

[33] Goswami DY. Principles of Solar Engineering. 3rd ed. Group LLC: Taylor & Francis; 2015

[34] Padilla RV, Fontalvo A, Demirkaya G, Martinez AQ, Arturo G. Exergy analysis of parabolic trough solar receiver. Applied Thermal Engineering. 2014;**67**:1-8. DOI: 10.1016/j.applthermaleng.2014.03.053

[35] Moran JM, Shapiro NH, Boettner DD, Bailey B. Fundamentals of Engineering Thermodynamics. 7th ed. USA: John Wiley and Sons; 2011

[36] Fujiwara M. Exergy analysis for the performance of solar collectors. Journal of Solar Energy Engineering. 1983;**105**: 163-167. DOI: 10.1115/1.3266360

[37] Petela R. Exergy of undiluted thermal radiation. Solar Energy. 2003; **74**:469-488. DOI: 10.1016/S0038-092X(03)00226-3

[38] Badescu V. How much work can be extracted from diluted solar radiation? Solar Energy. 2018;**170**:1095-1100. DOI: 10.1016/j.solener.2018.05.094

[39] Kumar D, Kumar S. Thermal performance of solar parabolic trough collector at variable flow rates: An experimental investigation. International Journal of Ambient Energy. 2018;**39**(1): 93-102. DOI: 10.1080/01430750.2016.1269673

[40] Marefati M, Mehrpooya M, Shafii MB. Optical and thermal analysis of a parabolic trough solar collector for production of thermal energy in different climates in Iran with comparison between the conventional nanofluids. Journal of Cleaner Production. 2018;**175**:294-313. DOI: 10.1016/j.jclepro.2017.12.080

[41] Liang H, You S, Zhang H. Comparison of different heat transfer models for parabolic trough solar collectors. Applied Energy. 2015;**148**: 105-114. DOI: 10.1016/j.apenergy.2015.03.059

Chapter 6

Perspective of NH_3 as a Clean Energy Zero-Carbon Fuel

Ali Raza, Marva Hadia, Zunaira-Tu-Zehra and Liaquat Ali Khan

Abstract

This research provides a comprehensive exploration of the potential of NH_3 as a transformative element in the pursuit of clean and sustainable energy solutions. With a growing global emphasis on decarbonization, the study delves into the unique attributes of NH_3 as a zero-carbon alternative fuel. The work examines NH_3's applications across various sectors, including transportation, industry, and power generation, both as a direct fuel and as a hydrogen precursor. Addressing the challenges associated with NH_3 production, the study highlights the ongoing research efforts to develop carbon-free production methods and ensure safe handling, providing a current snapshot of the field. By presenting a comprehensive analysis of the perspectives surrounding NH_3 as a clean energy alternative, this research contributes to the discourse on transitioning to a sustainable, low-carbon energy future. Amidst the pressing issue of climate change and the urgent need to transition to sustainable energy sources, NH_3 has emerged as a promising contender. This chapter thoroughly examines the multifaceted perspective of NH_3 as a clean energy, zero-carbon alternative fuel. It provides insights into its properties, production methods, current applications, challenges, and the transformative potential it holds for a greener future.

Keywords: fuels, NH_3, zero-carbon, renewable energy, diesel

1. Introduction

It has been crystal clear to everyone that energy is the most significant requirement for people and their existence. Energy consumption has constantly been increasing due to rising energy demands. New energy-intensive technologies, vehicles, systems, and applications are entering our daily routines every day, drastically increasing the fuel and power requirements for economic activities and societal developments. Such needs have impacted the energy equation with various environmental, health, and sustainable development constraints. It is now severely necessary to set up the energy equation without hydrocarbon fuels, and it is therefore fully recognized by many researchers, scientists, organizations, companies, etc. that it is time to move to renewables and clean fuels (particularly with hydrogen and NH_3) which have been advocated by Dincer [1] for many years. In a recent perspective article, [1] he declared that the COVID-19 coronavirus is closing the carbon age but opening the hydrogen age, and he considered the year 2020 a turning point that is even more necessary for human health and human welfare. In conjunction with this, there is a strong need to develop the technologies and economies to make a smooth transition as quickly as possible. That is why this perspective article takes NH_3 into consideration for a fair evaluation and suggestions for better combustion practices. The global energy landscape is transforming profoundly as societies grapple with the dual challenges of meeting rising energy demand and combating climate change. While abundant and affordable, traditional fossil fuels are also the primary drivers of greenhouse gas emissions, leading to detrimental environmental impacts. The search for clean, renewable, and sustainable energy alternatives has become imperative in this context [2–6].

One such alternative that has garnered significant attention in recent years is NH_3. While traditionally known as a critical component in fertilizers, NH_3 possesses attributes that make it an attractive candidate for clean energy applications. This paper explores the perspective of NH_3 as a zero-carbon alternative fuel, examining its potential to revolutionize various sectors and contribute to a low-carbon future. It is a well-known fact that almost one-third of the total consumed energy in the world is used in the transportation sector, where fossil fuels are primarily used to produce standard transportation fuels such as diesel, gasoline, and jet fuel. Their extensive use has been causing very high levels of greenhouse gases, ranging from 20% to 30%, depending on the Nation's development. Although manufacturers and governments have made significant attempts to transition to electric and hybrid vehicles, it seems that it is not possible to complete this transition in a short time due to infrastructure, economic, and raw material issues. However, the current environmental indicators indicate the requirement for quick and practical actions. Moreover, the diesel- and gasoline-powered generators used in residential, commercial, utility sectors, and off-grid applications contribute to fossil-based fuel consumption and increase CO_2 emissions. Using NH_3 in combustion processes, such as internal combustion engines and gas turbines, can be a crucial solution in a faster transition to the hydrogen economy [7, 8].

2. Understanding NH3

Ammonia, with its chemical formula NH_3, is a compound composed of one nitrogen atom bonded to three hydrogen atoms. It is colorless, pungent-smelling, and highly soluble in water. NH_3 is primarily produced through the Haber-Bosch process, which involves the reaction of nitrogen and hydrogen under high temperatures and pressure. Most NH_3 is used in fertilizers, but its applications extend far beyond agriculture.

One key attribute that makes NH_3 attractive as a clean energy alternative is its high energy density by volume. This allows for efficient storage and transportation, making it suitable for various energy applications. Additionally, it can be stored as a liquid at ambient temperature and moderate pressure, simplifying logistics compared to gaseous hydrogen.

3. Production methods

While the traditional Haber-Bosch process for NH_3 production is energy-intensive and reliant on fossil fuels, there is growing interest in alternative pathways that utilize renewable energy sources. Electrolysis of water to produce hydrogen and nitrogen capture from the air offers a pathway to green NH_3 production. Renewable energy sources such as solar and wind power can provide the electricity needed for electrolysis, resulting in zero-carbon NH_3 production.

Other innovative approaches include using green hydrogen derived from biomass or wastewater treatment processes. By leveraging organic waste streams, these methods produce NH_3 without carbon emissions and contribute to waste management and resource utilization.

Ammonia (NH_3) is recognized as a colorless gas with a sharp, pungent odor at room temperature, composed of nitrogen and hydrogen. Pure NH_3 is hygroscopic, meaning it readily absorbs moisture from the air and easily dissolves in water. Despite its corrosive nature due to its alkaline properties, NH_3 is one of the most widely produced industrial chemicals globally. Over 75% of the ammonia produced is utilized in agriculture as fertilizer. Additionally, NH_3 serves as a working fluid in refrigeration cycles and is a common ingredient in household cleaning products.

Figure 1 shows the global production levels of ammonia, with data from 1945 to 2017 depicted by a blue line, sourced from Reference [6]. In 2019, the worldwide production of NH_3 reached approximately 146 million metric tons. Only a small fraction (around 4%) of NH_3 is used directly, while the majority is employed as a chemical in industrial applications or as fertilizer. With increasing environmental concerns and efforts to reduce CO_2 emissions, the use of NH_3 in the energy sector is anticipated to grow. Various initiatives are underway to adopt NH_3 as a carbon-free fuel, with numerous NH_3-related projects announced for power generation, off-grid applications, and internal combustion engines.

Moreover, the demand for NH_3 is expected to rise with the proliferation of hydrogen-powered systems, as NH_3 is an effective hydrogen carrier. We project a significant increase in the production and consumption of NH_3, as illustrated by the red-dotted line in **Figure 1**. By 2050, it is estimated that global production of NH_3 will reach about 1.2 billion metric tons, an increase of nearly 8.2 times the amount produced in 2019 (**Figure 2**) [1, 9].

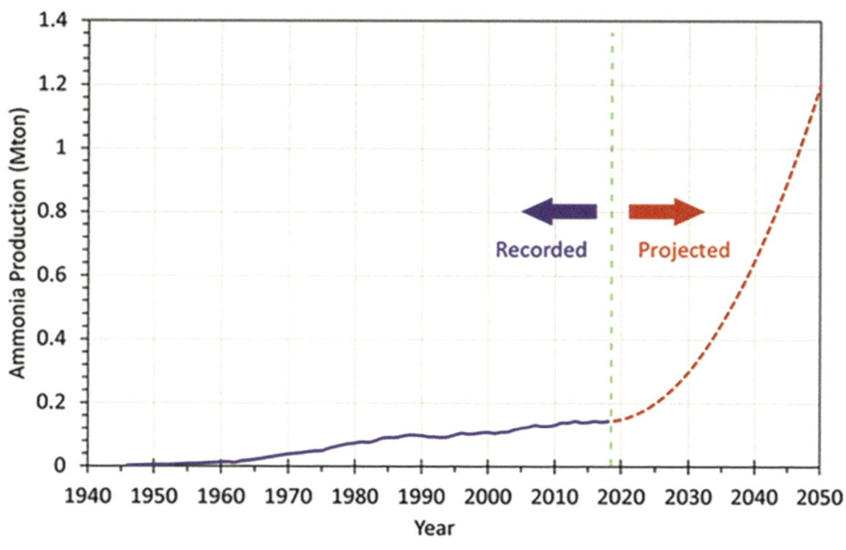

Figure 1.
The global production of NH_3 [1].

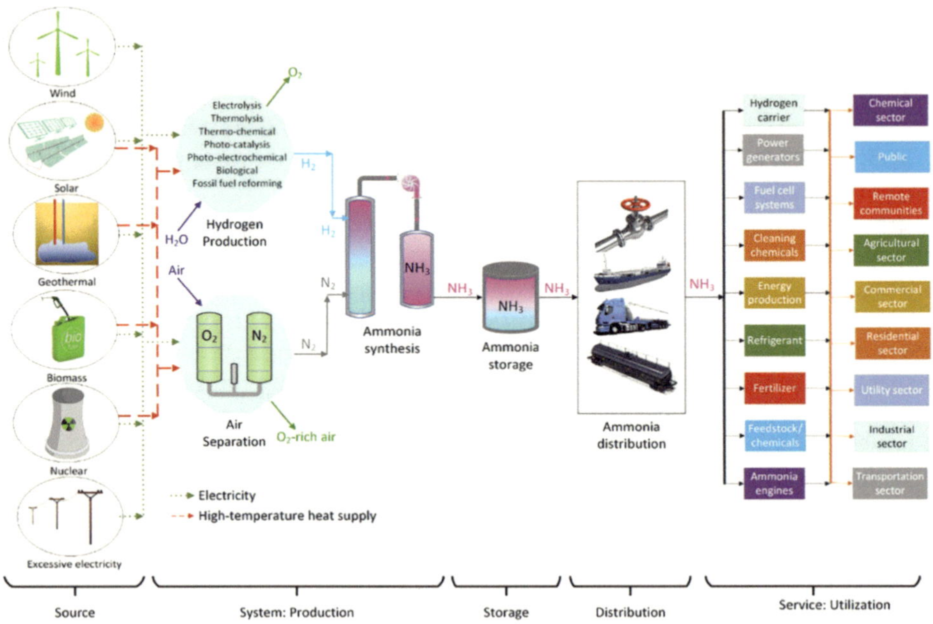

Figure 2.
An economic cycle of NH_3 from production to utilization [1].

4. Current applications and challenges

NH_3 has a wide range of applications beyond agriculture, including refrigeration, cleaning agents, and the manufacturing of pharmaceuticals and explosives. However, its potential as a clean energy fuel is still in the early stages of exploration. One of the main challenges facing the widespread adoption of NH_3 as a fuel is the lack of infrastructure for production, distribution, and utilization.

Additionally, safety concerns related to the handling and storage of NH_3 must be addressed through robust regulations and risk management practices. Furthermore, the economic viability of green NH_3 production remains a challenge, as it currently requires significant investment and may not yet be cost-competitive with conventional NH_3 production methods.

5. The promise of NH3 as a zero-carbon fuel

Despite these challenges, NH_3 holds excellent promise as a zero-carbon fuel with the potential to decarbonize various sectors of the economy. NH_3-powered fuel cells offer an alternative to internal combustion engines in transportation, providing clean and efficient propulsion for vehicles ranging from cars to ships and aircraft.

Moreover, NH_3 can be used as a fuel in power generation, either in combined-cycle power plants or co-fired with coal in existing facilities. With the option of capturing and sequestering CO_2 emissions, NH_3-based power generation can achieve zero net carbon emissions, contributing to climate mitigation efforts [10–12].

6. Challenges with NH3 combustion

It is essential to point out that NH_3, when used for combustion, has some side effects, just like all drugs have some side effects. Such side effects (so-called challenges) may be listed as follows:

- High ignition temperature
- Low flame velocity
- Slow chemical kinetics

Since increasing research efforts are being made to minimize the impacts and improve combustion performance, some essential remedies to overcome the issues are necessary. These will be discussed in the following subsection.

7. Challenges and driving innovation

Concerted efforts are needed to overcome technical, economic, and regulatory challenges and realize NH_3s promise as a zero-carbon fuel. Research and development initiatives focused on improving production efficiency, enhancing safety standards, and reducing costs are crucial.

Collaboration between governments, industries, and research institutions is essential to accelerate innovation and increase NH_3 production and utilization. Policy incentives, carbon pricing mechanisms, and international cooperation can further incentivize investment in sustainable NH_3 technologies.

8. Conclusion

In conclusion, NH_3 represents a compelling opportunity to transition toward a sustainable energy future. Its unique properties, coupled with advances in production methods and utilization technologies, position it as a versatile and scalable solution for decarbonizing various sectors of the economy.

By addressing the challenges and seizing the opportunities presented by NH_3, we can accelerate the transition to a low-carbon energy landscape and mitigate the impacts of climate change. With concerted efforts from stakeholders across the globe, NH_3 has the potential to play a pivotal role in shaping a cleaner, greener, and more resilient future for generations to come.

Conflict of interest

The authors declare no conflict of interest.

Notes/thanks/other declarations

The authors thank the National University of Technology (NUTECH) for their support.

Glossary

- *NH_3*: ammonia, a colorless, pungent-smelling compound composed of one nitrogen atom bonded to three hydrogen atoms, renowned for its high solubility in water.
- *Decarbonization*: the gradual reduction of carbon dioxide emissions, especially from human activities such as burning fossil fuels.
- *Hydrogen economy*: an economic system reliant on hydrogen as the primary energy carrier, often sourced from renewables to minimize carbon emissions.
- *Green hydrogen*: when hydrogen is generated via electrolysis using renewable energy, it produces energy without any emissions.
- *Zero-carbon fuel*: a fuel source producing no net carbon dioxide emissions when both produced and consumed, aiding in the fight against climate change.
- *Haber-Bosch process*: a chemical reaction process that produces ammonia from nitrogen and hydrogen, typically involving high temperatures and pressure.
- *Electrolysis*: a chemical process utilizing electricity to split water into hydrogen and oxygen gases, commonly employed in green hydrogen production.
- *Carbon pricing mechanisms*: economic policies imposing a price on carbon dioxide emissions to encourage emission reductions.
- *Combined-cycle power plants*: facilities utilizing gas and steam turbines to generate electricity, enhancing overall efficiency.
- *Fuel cells*: electrochemical devices convert fuel chemical energy directly into electricity and heat.
- *Internal combustion engines*: these engines generate mechanical power by burning fuel within a combustion chamber. They are widely used in transportation and power generation.
- *CO_2 capture and sequestration*: this is the process of capturing carbon dioxide emissions from industrial processes or power plants and storing them underground to prevent release into the atmosphere.
- *Hygroscopic*: a property of substances, like ammonia, that absorb moisture from the air.
- *Alkaline properties*: properties making a substance basic or alkaline, usually characterized by a pH greater than 7.
- *Fossil-based fuel*: fuel derived from fossilized organic materials, emitting carbon dioxide when burned.
- *Renewable energy sources*: naturally replenished energy sources like sunlight, wind, and biomass, emit no greenhouse gases during operation.

- *Waste management*: processes involving collection, transportation, recycling, and disposal of waste to reduce environmental pollution and conserve resources.
- *Resource utilization*: the efficient use and management of natural resources to meet human needs while minimizing environmental impact.
- *Risk management practices*: strategies for identifying, assessing, and mitigating potential risks associated with specific activities or processes.
- *Stakeholders*: individuals, groups, or organizations with an interest or concern in a particular issue, project, or outcome, including government agencies, businesses, communities, and non-profit organizations [13–15].

Author details

Ali Raza[1*], Marva Hadia[2], Zunaira-Tu-Zehra[1] and Liaquat Ali Khan[1]

1 National University of Technology (NUTECH), Islamabad, Pakistan

2 National University of Sciences and Technology (NUST), Islamabad, Pakistan

*Address all correspondence to: ali.raza@nutech.edu.pk

References

[1] Erdemir D, Dincer I. A perspective on the use of ammonia as a clean fuel: Challenges and solutions. International Journal of Energy Research. 2021;**45**(4):4827-4834

[2] Dincer I. Covid-19 coronavirus: Closing carbon age, but opening hydrogen age. International Journal of Energy Research. 2020;**44**(8):6093

[3] Wood E. California Opens "Diesel Alternative" Discussion in Microgrid Proceeding. 2020. Available from: https://www.microgridknowledge.com/californiamicrogrid-19-09-009/ [Accessed: September 19, 2020]

[4] Raza A, Miran S, Islam TU, Malik KI, Hadia M. Numerical study of evaporation modelling for different fuels at high operating conditions in a diesel engine. Engineering Proceedings. 2021;**12**(1):8

[5] Times F. Japan Bets on Ammonia as the Fuel of the Future. London, UK: Financial Times; 2020

[6] Rasmussen V. AMMONIA-Fueled Gas Turbine Power Generation. 2017; n.d. Available from: https://www.AMMONIAenergy.org/articles/AMMONIA-fueledgas-turbine-power-generation/#:text=HideakiKobayashi %2C professor at the,AMMONIA in a gas turbine.&text=But because AMMONIA burns without,in a low-carbon society [Accessed: September 19, 2020]

[7] Brown T. Japan's NYK and Partners to Develop Ammonia Fueled and Fueling Vessels. VA, USA: Ammonia Energy Association

[8] Raza A, Khurram M, Khan MA, Durez A, Khan LA. Comparison of evaporation in conventional diesel and bio-fuel droplets in engine cylinder. In: Exergy-New Technologies and Applications. London, UK: IntechOpen; 2023. pp. 167-169

[9] U.S. Geological Survey. AMMONIA statistics. In: Kelly TD, Matos GR, editors. Historical Statistics for Mineral and Material Commodities in the United States. Reston, VA; 2016

[10] Mishra DP, Rahman A. An experimental study of flammability limits of LPG/air mixtures. Fuel. 2003;**82**(7):863-866

[11] Ramasamy D, Kadirgama K, Rahman MM, Zainal ZA. Analysis of compressed natural gas burn rate and flame propagation on a sub-compact vehicle engine. International Journal of Automotive & Mechanical Engineering. 2015;**1**:11

[12] Portal E. The Alternative Fuels Data Center–Providing Biofuels Tools and Information. U. S. Department of Energy. n.d. Available from: www.eere.energy.gov/afdc [Accessed: October 21, 2020]

[13] Raza A, Mehboob H, Miran S, Arif W, Rizvi SF. Investigation on the characteristics of biodiesel droplets in the engine cylinder. Energies. 2020;**13**(14):3637

[14] Raza A, Miran S, Khurram M, Islam TU, Hadia M. Numerical investigation of evaporation modelling for different diesel fuels at high temperature and pressure in diesel engine. Journal of Engineering Research. ICEPE Special Issue. 2022. pp. 1-15

[15] Dincer I, Acar C. Review and evaluation of hydrogen production methods for better sustainability. International Journal of Hydrogen Energy. 2015;**40**(34):11094-11111

Chapter 7

The Svedberg Number, 1.54, as the Basis of a State Function Describing the Evolution of Turbulence and Dispersion

Alfredo Constaín Aragón

Abstract

For natural flows, a definition equivalent to the Chezy-Manning mechanical equation is developed, but based on Coulomb interactions, which includes a state function that describes the evolution of tracer particles in turbulence. This makes it possible to overcome reductionist approaches (Navier-Stokes), which are limited by their complexity. This function shows the variation of the degrees of freedom of dispersion, as well as the statistical coupling of the solute with the flow, allowing to characterize very large or complex channels in "Dynamic Equilibrium." An approach is also developed that links this function to universal constants of fractal motion.

Keywords: turbulence, linear regime, irreversibility, exergy, state functions

1. Introduction

1.1 Definition of the "statistical sufficiency" of molecular phenomena

According to A. Annila et al. [1, 2] molecular systems at the macroscopic level, the statistical distribution of energy in these systems unfolds around the average molar thermal energy, R^*To, (R as the gas constant, and To as the ambient Kelvin temperature) as shown in **Figure 1**.

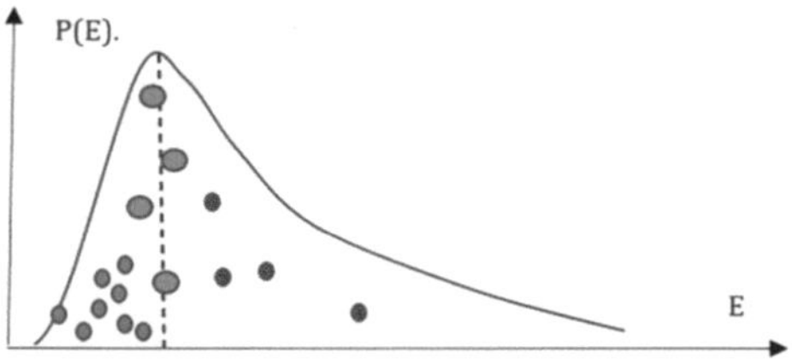

Figure 1.
Energy distribution on a molecular system.

In "steady state," (time invariant) oscillating particles are grouped into quantized energy zone (h^*f) according to their probability. For the low-energy zone ($E < RTo$), many oscillators (blue) are grouped together but with low energy per unit; for the high-energy zone ($E > RTo$), only a few oscillators (red) will be able to supply the required high energy. In the central zone ($E<RTo$), close to the average value, there are enough oscillators with medium energy (green) in such a way that the sum of which is greater than in the low and high zones, with the highest probability.

In an open system, it is said to be "statistically sufficient" when the fluctuations around the mean (most likely) value are small, such that it does not change significantly, which is achieved if the "activation energy" is much lower than the average energy, $W < <RTo$.

1.2 Local equilibrium and "statistical sufficiency": water fluidity and turbulent phenomena

The concept of local equilibrium in systems in non-thermodynamic equilibrium is essential for applying the usual thermodynamic equations to these systems, in which, despite not having "well-defined" variables such as temperature, energy, entropy, etc. there are well-defined "densities" of these variables, making it possible to analyze them consistently [3].

The validity of local equilibrium is limited by the ease with which dynamic processes restore "well-defined" initial distributions, in presence of fluctuations. In continuum mechanics, e.g., in hydrodynamics, when fluctuations act and the system tries to return to its "local" equilibrium, if the system in its dynamics suffers a nonlinearity, e.g., the term $v^*grad\ (v)$ in the *CM* equation, the return to the previous point of local equilibrium is not possible because the results of the equation are not univocal, diversifying the behavior of the system. This effect of positive feedback is at the root of the problem of the "non-computability" of nonlinear phenomena (**Figure 2**) [4].

The range of validity of the local equilibrium in certain phenomena is not generally known from a microscopic perspective. However, it can be said that for gases, liquids, and transport phenomena, it exists well enough if the gradients of the intensive state variables are small and vary very slowly.

According to J. Frenkel in his "Kinetic theory of liquids" [5], due to the high *fluidity* properties of water—i.e., the capability of yielding a stress—this liquid allows for shear movements with almost no resistance. This behavior of water is measured by the ratio of the time, "τ" required to construct a population of "holes" in the lattice of liquid, to the basic period of atomic oscillation, $\tau_o \approx 10^{-13}\ (sec)$.

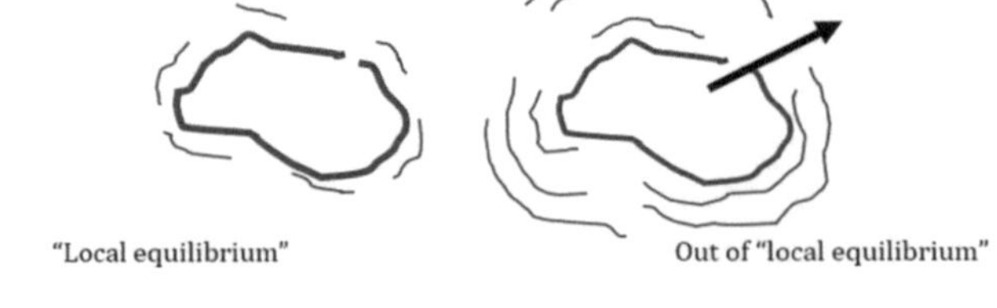

Figure 2.
Local equilibrium and fluctuations.

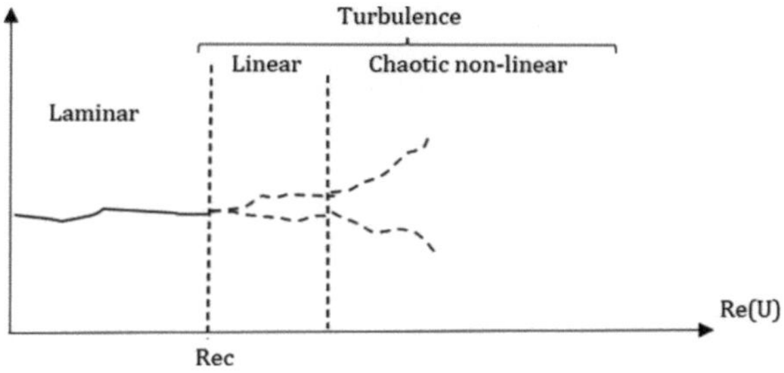

Figure 3.
Regimes of thermodynamics.

$$\frac{\tau}{\tau o} \approx e^{\frac{\Delta W}{RT}} \tag{1}$$

This time τ is named the "half-life" of a particle in a "vacancy" position (hole) in the lattice of the liquid. If the ratio $\tau/\tau o$ is small, the fluidity of the liquid will be high, and if it is large, it will be low. Here, ΔW is the so-called "activation energy" in fluid, $\Delta W \approx 0.58$ *KJ/mol*, being currently less than the maximum work of "forming a gap" in the water lattice, so, if the thermal heat is $RT \approx 2.5$ *KJ/mol (at 300 K)*, then $\tau/\tau o \approx 1.26$, i.e., enough close to the ideal value of fluidity [3].

To define its state of "turbulence" in the case of water, the so-called "Reynolds Number" (*Re*) is used, which is the ratio between the product of a characteristic length, *L*, and the average velocity of the flow, *U*, and the "kinematic viscosity", ν, which are dimensionally equivalent.

$$Re = \frac{U * L}{\nu} \tag{2}$$

The turbulence begins when *U* threshold is transposed, called the "critical" Reynolds Number (*Rec*), when nonlinearity also begins to occur at the microscopic level, but maintaining approximately the nature of "statistical sufficiency," in which thermodynamic *force* (gradient of potential energy) quickly generates a flow (mass transport), without giving time for nonlinear phenomena to fully develop, thanks to its high fluidity.

Then, fluctuations, not so large, do not appreciably modify the energetic characteristic of the system. Water in turbulence, with velocities in a low and medium range, is "quite statistical," i.e., the energy changes that occur will not significantly change the quasi-equilibrium condition, falling into the "Linear" regime of irreversible thermodynamics, in which the system evolves "slowly," and the state remains stable (**Figure 3**).

2. Geomorphology and flow: factors that determine the behavior of natural streams, and the average velocity of the flow determines two important aspects of natural channels

2.1 Average velocity and geomorphology

The complex behavior of natural channels is marked by the role of flow velocity, both in the configuration of the geomorphology (profiles, geometry, and nature of the

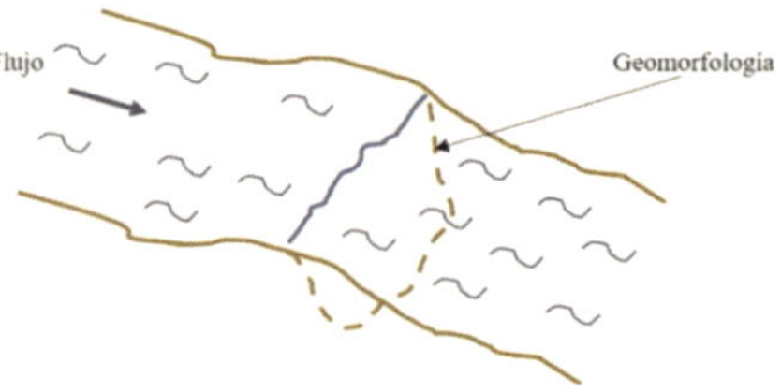

Figure 4.
Interaction of flow and geomorphology.

riverbed) and in the dynamics of the flow itself (distributions and values of the velocity vectors) [6]. This deep relationship is shown in the law of Conservation of Flow, in which the flow rate, Q, depends on the mean velocity, U, and the cross-sectional area of the flow, Ayz (**Figure 4**).

$$Q \approx \frac{U}{Ayz} \tag{3}$$

2.2 Average velocity and flow behavior

The average velocity not only interacts with the geomorphology but also with the flow itself. This is observed in the basic differential equation of point velocity, v, of NS equation [7], which depends on a nonlinear way of pressure, p, temperature, T, external forces, f, water density, ρ, and time.

$$\rho\left(\frac{\partial v}{\partial t} + v * \nabla v\right) = -\nabla p + \nabla * T + f \tag{4}$$

When the velocity value passes a threshold (Rec), the fluctuations by shear start to be amplified by nonlinearity and turbulence appears [8]. In this condition, the cross-sectional distribution of the longitudinal point velocity vectors is configured as a "Random Variable," mainly described by the normal distribution, since the viscous rupture of the flows can be modeled as a bifurcation with two probable, symmetrical values, which are congruent with the measurement errors of the Gaussian bell [7].

This homogeneity in the results is reflected in the classical (Pr) distribution, which is essentially "flat" except at the edges due to the progressive effect of the viscosity of the boundary (**Figure 5**).

Figure 5.
Prandtl distribution of velocities.

2.3 Thermodynamic forces and flows of natural streams in the "linear" regime of irreversible thermodynamics. Convergence with the CM equation

Turbulent dynamics are composed of heterogeneous movements, usually rotational in nature, to expend free energy in the fastest way [9]. In the first stage, the strength of these movements is diminished by the great fluidity of the water, due to its nature of "statistical sufficiency."

Then, for certain "normal" low range of velocities, this condition fundamentally allows turbulent flows to develop in the "linear" region of irreversible thermodynamics where thermodynamic "forces" (potential gradients) are linearly related with thermodynamic "flows" (mass or charge transport). This condition is examined for a natural flow as follows:

From the definitions of thermodynamic forces and flows, the following expressions can be established for natural flows, with ρ the water density, and U the mean velocity, obtaining the flow, J:

$$J \approx \frac{\frac{\Delta m}{Ayz}}{\Delta t} \approx \frac{\frac{\Delta m}{\Delta Y * \Delta Z}}{\Delta t} * \frac{\Delta X}{\Delta X} \approx \frac{\left(\frac{\Delta m}{\Delta V}\right) * \Delta X}{\Delta t} \approx \rho * U \tag{5}$$

For the Force, F, it holds, considering the potential energy gradient:

$$F \approx \Delta m * g * \Delta h \approx \frac{(\Delta m * g * \Delta h) * \Delta X}{\Delta X} \approx \Delta m * g * \frac{\Delta h}{\Delta X} * \Delta X \approx \rho * Ayz * g * S \tag{6}$$

Considering the basic relation of linear thermodynamics of irreversibility, force as cause and flow as a consequence:

$$J \approx L * F \tag{7}$$

Therefore, from (4), (5), and (6), in this case, the *mean velocity of flow is proportional to the Slope S.*, if valid "linear" regime:

$$U \approx L * (Ayz * g) * S \tag{8}$$

It is now necessary to analyze the classical CM equation of flow in order to compare it with the previous result, in which "R" is the hydraulic radius, "n" is the opposition to the flow (Roughness), and "S" is the slope of the flow.

$$U \approx \frac{R^{\frac{2}{3}}}{n} \sqrt{S} \tag{9}$$

It is obvious that the mean velocity is not linear with the slope, but proportional to the *square root of it.* Now, it is interesting to establish the degree of "non-linearity" of CM Eq. (9) with respect to the thermodynamic Eq. (8) and compare the curves corresponding to each case (**Figure 6**).

By separating the range of the slope into segments to compare $\sqrt{S}$ with S, it can be verified that the blue curve ($\sqrt{S}$) can be linearized by lines that fit in each range, from 0–10%, with an approximate relative error of 10% in larger segment.

This result is compatible with a McLaurin-Taylor (Tl) series expansion for representative values [10]. So, it can be said that the classical equation of hydraulics corresponds well with a linear approximation on the slope and that therefore in the

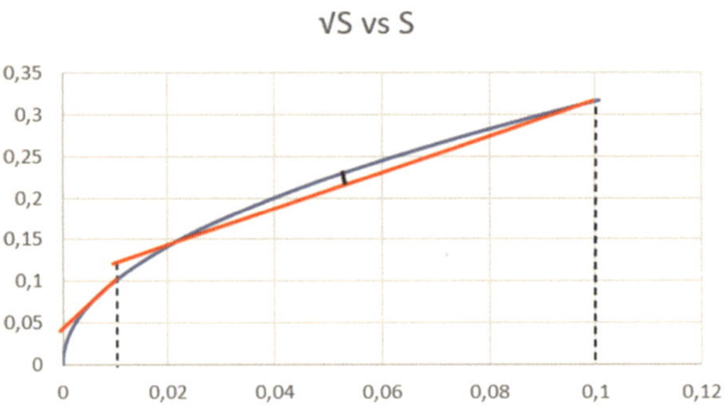

Figure 6.
Linear fitting of Chezy-manning equation.

practical range of the slopes, the condition of thermodynamic "linearity" of the flow is approximately satisfied, compatible with the hypotheses proposed in this article. This may be called the "linearity approximation."

2.4 The linear regime of irreversible thermodynamics and minimum entropy production. Effect of the "linearity approximation"

According to I. Prigogine, when both Forces and thermodynamic flows can be defined for an open system, their product is equal to the entropy production per unit volume, $\Delta iS/(dt^*dV)$. [3], and also that the entropy, S, has a maximum (**Figure 7**).

Consistent with this situation, the distribution of probabilities in the volumes of the flow (according to Boltzmann's concept) ideally tends to be a well-extended "coarse-grained" volume, [11] in which the microstates are indistinguishable. But actually, it is not "ideal," considering that there is a (small) error for the "linearity approximation." This special circumstance makes the "coarse-grained" areas have disseminated "gaps" that are detected as groups of more vigorous or more numerous fluctuations, showing the "intermittencies" that are observed experimentally in real flows (**Figure 8**).

So, in the real case of turbulent water flow, it will be shown that the *fluidity* is low, but it is not zero, then it is necessary to consider the existence of "intermittences," leading the "coarse-grained" regions painted with "patches" that are not representative of the ensemble, but being concordant statistically with PrN distribution. In this case, it is necessary to consider the velocity "Variance," σ_u, that involves the deviations from the "linearity" of the turbulent phenomenon. An expression of measuring this parameter is proposed below (**Figure 9**).

$$U(measured) \approx < U > \pm \sigma_U \tag{10}$$

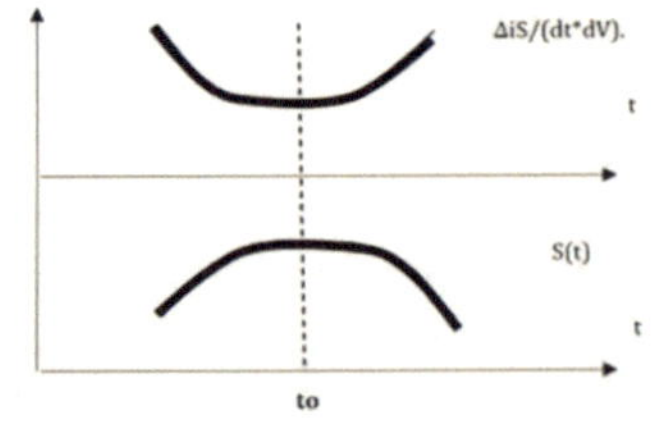

Figure 7.
Behavior of entropy and entropy production.

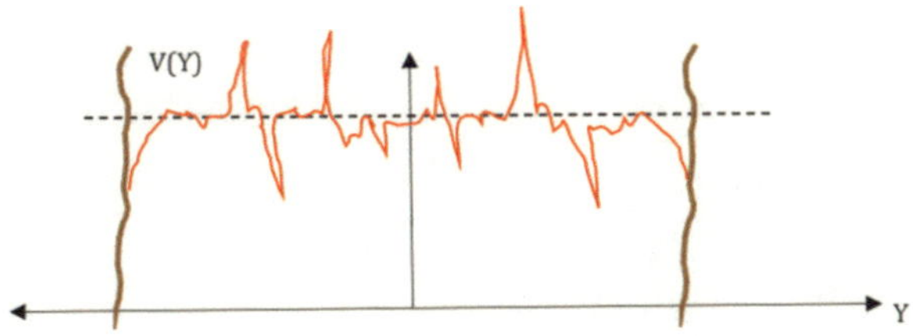

Figure 8.
Effect of nonlinear fluctuations.

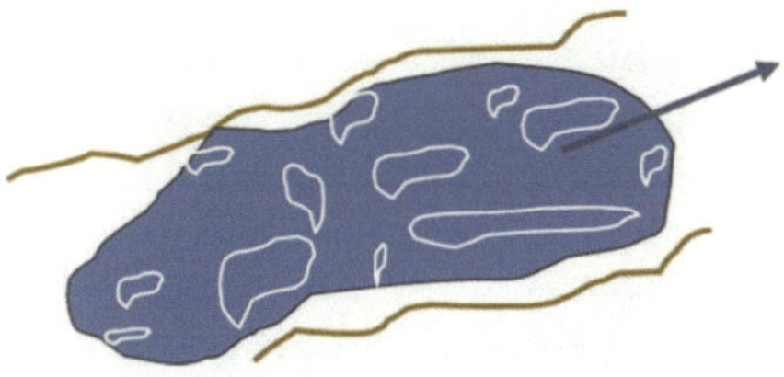

Figure 9.
Variance as effect of non-linearities in flow.

This parameter, σ_U, must be calculated not only in terms of purely hydraulic magnitudes since the dynamic equilibrium condition depends on a larger set of variables, which influence variance.

3. "Dynamic equilibrium" in the channels and "linear" region of the thermodynamics of irreversible processes

The linear region of irreversible thermodynamics is characterized by the fact that open systems that evolve in this regime ideally present "stable states," that is, they do not vary with time, while their entropy production is a minimum, and thermodynamic flows tend to be constant, as is actually observed in natural waterways.

3.1 Basic characteristics of dynamic equilibrium in natural streams

The processes in natural turbulent streams establish a tendency for the flow to oscillate around a certain value of the flow state parameters, which is called "Dynamic equilibrium," in that locally there are variations, but its state tends to stabilize (**Figure 10**).

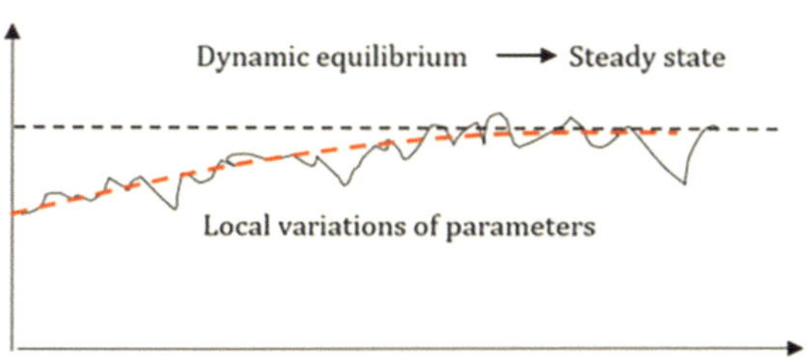

Figure 10.
Dynamic equilibrium tends to steady state in flows.

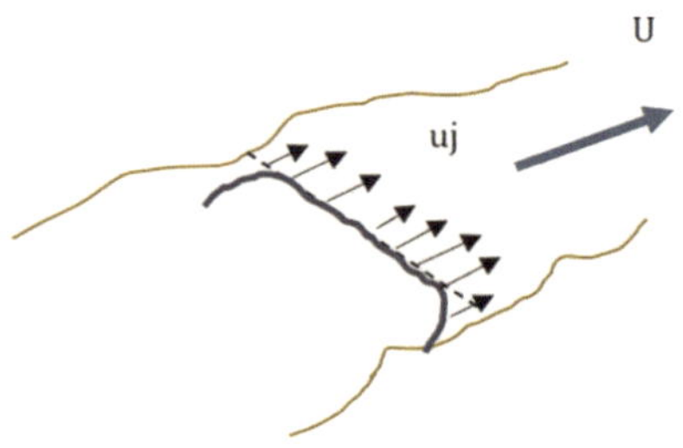

Figure 11.
Local and general velocities in steady state.

Local variations are caused by contrary forces, and the point values are such that the energy and its distribution in its course take the most probable configurations, responding in detail to the following general criteria [12].

A. If river systems are in a "steady state" (invariant with time) due to the self-adjusting action of opposing forces, then the entropy of the open system, *S*, is the maximum possible, in congruence with the constraints imposed by the environment, and therefore the internal production of entropy, $\Delta iS/(dt)$, is the minimum possible.

B. At "dynamic equilibrium," if the entropy production in the flow is the minimum possible and almost constant, the probabilities in the flow realm are ideally equivalent, this implies that the mass transport rates are ideally equivalent to each other, and concordant with the overall mean transport rate, leading to the fulfillment of the Prandlt distribution for the longitudinal velocities on the *transverse axis* of a river (**Figure 11**).

$$uj \approx uk \forall j \text{ and } k\ U \approx \frac{\sum_n u(n)}{n} \tag{11}$$

The mean velocity is almost constant in all its course, as found by L. Leopold. Becoming very similar at birth to your delta, this is:

u1<u2<u3<u4<u5<u6<u7<u8<u9<u10 (**Figure 12**) [13].

C. Under these conditions, the entropy production is equal to the product of the thermodynamic force and flow, with constant temperature and pressure, and no electric or magnetic fields.

$$Ps \approx \frac{\Delta iS}{\Delta t} \approx L * J * F \tag{12}$$

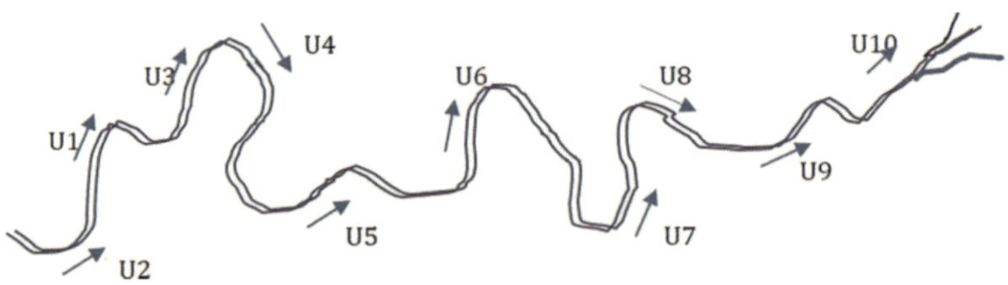

Figure 12.
Leopold's longitudinal continuity of velocity.

In 2.2, *F*, the "Force," is a gravitational gradient and *J*, the "Flow," is the mass transport per unit of cross-sectional area and time.

4. Formation of the tracer plume in a turbulent flow, and models that represent its dynamic

4.1 Thermodynamic foundation of the phenomenon

Tracers are salt inks that solubilize in water, with a very limited environmental impact, presenting physical or chemical properties that allow their detection by instruments, such as their fluorescence or conductivity, which enables their widespread use for the scientific characterization of flows [14].

An injection of tracer into the flow configures the creation of a "plume" that advances as an open, isothermal-isobaric system, which breaks the chemical equilibrium in the system, implying in some way the reorganization of the internal molecular distributions.

When the solute is injected into the flow, two phases take place: A. SOLVATION in which the dipoles of the water act on the ions of the salt, separating them from the crystal structure, that is, working against the forces of interaction of the crystal lattice. B. DILUTION, which happens once the solvation of all the ions has been completed, and they are incorporated into the flow, moving with their own speed that depends on the concentration of the salt, which varies over time (**Figure 13**).

The initial phase of SOLVATION is explained as the creation of hydrated ions (surrounded by water molecules), which must first be separated from the crystal lattice of the solid, and then electrically bonded to these molecules. This occurs because the energy of interaction between ions and dipoles destroys the crystal lattice.

This phase of the process, which is exothermic, decreases the potential energy of the whole so that the different particles are "ordered" better, which causes the entropy to decrease, which is expelled as heat to the outside.

This process may be interpreted as the battle and equalization of two opposing energies: the heat (enthalpy) of formation of crystal, and the heat of Solvation, given from the surrounding water (H_20), whose two hydrogen atoms and one oxygen atom act as separate electric charges (dipoles), whose specific interactions are responsible for the destruction of the solid, giving ions.

The enthalpy of formation (positive), $+\Delta Uo$, of the crystal structure of the solid, and the *Solvation heat* (negative), $-\Delta Hh$, which is added to the salt from the water, its

Figure 13.
Dye tracer salt dilution in streams.

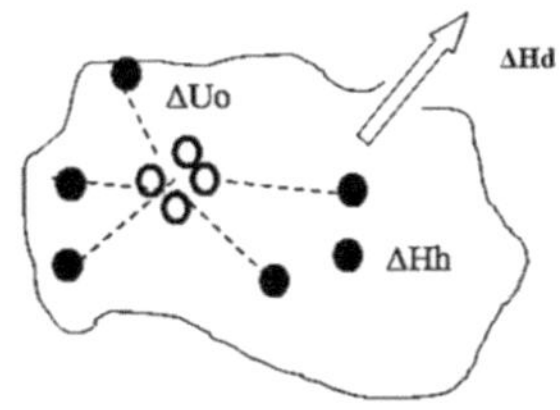

Figure 14.
Entropy expulsion in tracer system.

difference is the *Dilution heat*: $T^{*}\Delta Sd \approx \Delta Hd$, which balance equation is as follow (**Figure 14**).

$$\Delta Hd \approx T * \Delta iS \approx \Delta Uo - \Delta Hh \tag{13}$$

Thus, the joint action of ΔHh and ΔUo is spontaneous "internal" reorganizations in the system, which does not consume energy other than that shown in Eq. (13).

The next phase, called the DILUTION Phase, occurs when the generation of ions has ended when the solid structure of the solute is completely destroyed. At this point, it is necessary to consider the other interaction between the ions themselves, already in freedom, which are distributed forming an "ion cloud" that is a function of the Concentration, $C(t)$ (**Figure 15**).

This phase, like the first one, involves a decrease in the potential energy of the whole, then it is also exothermic, with the expulsion of a spreading heat, ΔHd, which is now proportional to the square root of the concentration, $\sqrt{C}$, which decreases over time, as established in electrochemical thermodynamics (**Figure 16**).

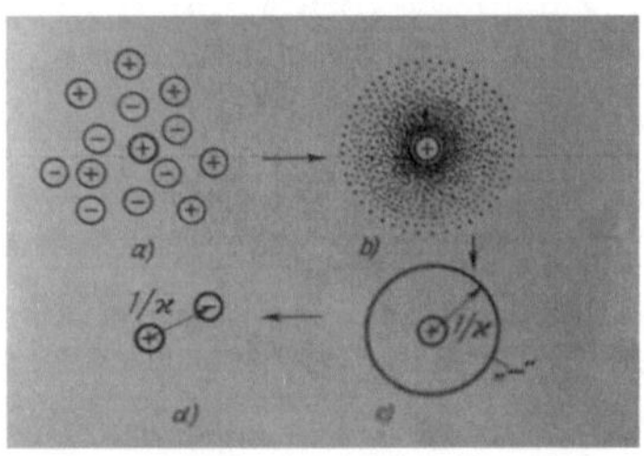

Figure 15.
Ion-ion interaction in tracer spreading [14].

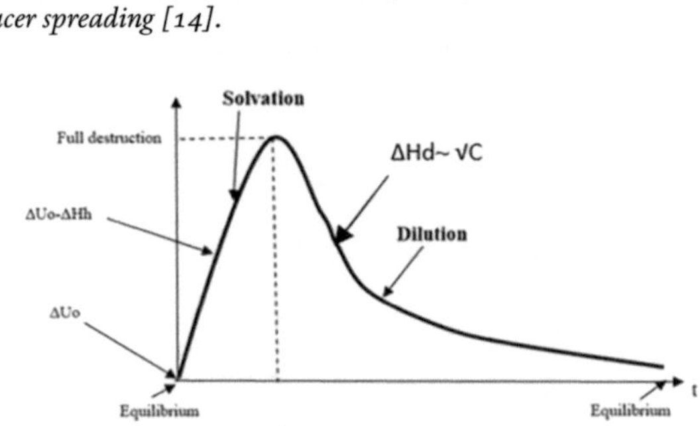

Figure 16.
Thermodynamic process of a tracer in flow.

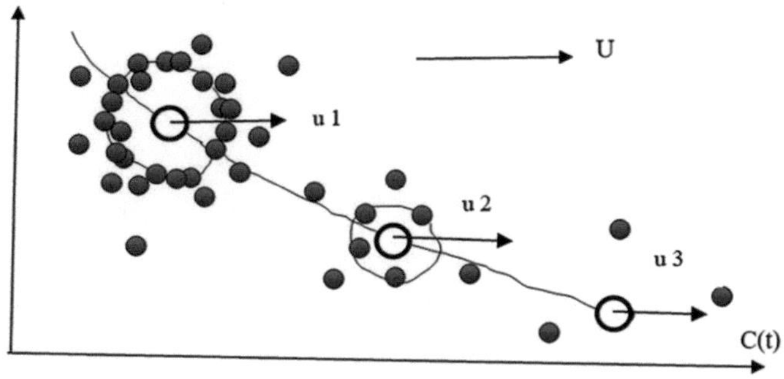

Figure 17.
Concentration affects the tracer velocity.

The interactions of the ions with each other reduce their mobility, which implies a "slowing down" effect on their movement. This effect is similar to the increase in friction by the Stokes mechanism.

Thus, for initial conditions (high concentration), the aggregate ions will have a small Stokes velocity, *u1*, for a medium concentration, they will have a higher velocity, *u2*, and in their final stage (low concentrations) their eigen velocity, *u3*, will be larger, equaling the velocity of the flow (**Figure 17**).

$$u1 < u2 < u3 \tag{14}$$

And:

$$u3 \approx U \tag{15}$$

This indicates that the transport time, *ts*, of the solute ions, depending on the decreasing concentration, will decrease as the effective velocity increases: *ts1 > ts2 > ts3*.

4.2 Advection-dispersion transport model

Although the transport of conservative substances in turbulent flows has been characterized by Fick's equation, initially developed to represent diffusive motions associated with Brownian movement, G.I [15, 16]. Taylor postulated in 1954 that the main mechanism for the mixing of solutes in the flow is the "shear" effect, which can be represented also by these equations. In other words, this statistical movement is proportional to the concentration gradient. This effect corresponds to the longitudinal breakdown of the minimum parcels of the flow by the action of "separation" by the point velocity vectors, which are a random quantity.

This model is important in that it shows that the mean velocity, *U*, is composed of an infinite number of different components, that this velocity value is a statistical result of all its components, and that the magnitude of the DISPERSION (Longitudinal Coefficient of Dispersion, *E*) depends on the magnitude of the velocity.

$$E \approx f\left(U, \frac{\partial C}{\partial X}\right) \tag{16}$$

While LONGITUDINAL DISPERSION (lengthwise) is essentially determined by the shear effect of velocity, TRANSVERSE DIFFUSION (across width) is basically

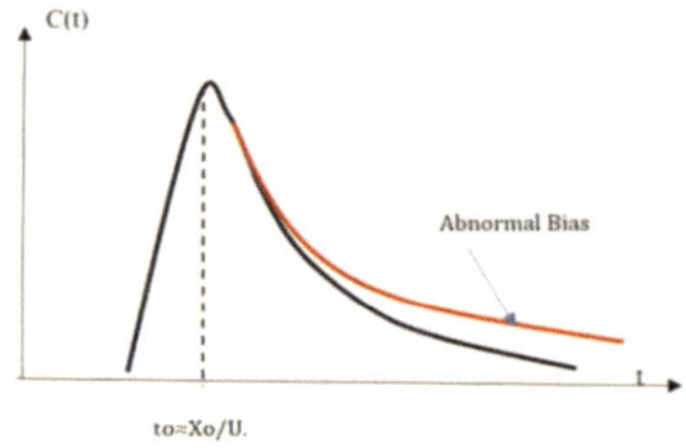

Figure 18.
"Abnormal" biased tracer curve.

determined by turbulent mixing. This and other efforts to characterize the mixture of solutes in turbulence aim to describe and explain the curve in time of the tracer's passage at a fixed point (Eulerian observer). Gathering all the knowledge, the equation of this curve has been established, according to the advection-dispersion model, which includes the average velocity and concentration in the cross-section of the flow, as functions of time and distance, *Tl* equation:

$$\frac{\partial C}{\partial t} + U\frac{\partial C}{\partial X} \approx E\frac{\partial^2 C}{\partial X^2} \tag{17}$$

The solution of which is Fick's equation, where *Xo* is the distance from sudden injection, *M* is the injected mass, and A_{yz} is the cross-section area.:

$$C(t, X) \approx \frac{M}{Ayz\sqrt{4Et}} e^{-\frac{(Xo-U*t)^2}{4Et}} \tag{18}$$

Numerous controversies have arisen regarding the shape of the bell-shaped (Gaussian) curve of the tracer's response, since although it maintains a bias due to the Galilean composition of the velocity *(Xo-U*t)* that exists in the Neperian function, many times the experimental bias observed is greater than expected (Abnormal bias), and some discrepancies about the peak concentration (**Figure 18**).

4.3 Transient storage (TS) and aggregated dead zone (ADZ) model for tracer evolution in flow

The TS Model has been proposed extending the classical advection-dispersion model but has been unable to adequately describe the abnormal bias in tracer curves [17, 18]. It assumes that there are "Storage zones" or areas that retain or trap groups of tracer particles on the periphery of the channel, which are subsequently released, affecting the longitudinal dispersion of the stream. The "Storage zones" are characterized by a proper concentration, *Cs*, where the solute is homogeneously distributed, and a volume defined by an area *AS* in the cross-section, while the central channel shows a concentration, *Cb*, and a volume defined by an area *Ab*. Between the two volumes, perimeter dead zone, and central channel, there is an exchange of particles, as already stated, defined by "α." They are not considered chemical reactions or non-conservative changes in the solute (**Figure 19**).

The objective of the model is to calculate parameters, such as: *As, Ab*, α, and the longitudinal coefficient of dispersion, *E*, that can reasonably describe the actual tracer curve evolution, with its abnormal bias and peak concentration value.

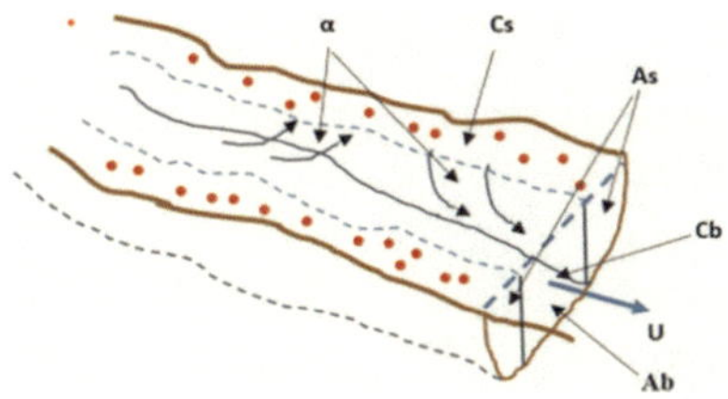

Figure 19.
Dead zone storing tracer in borders.

An initial difficulty is that the model's own equations (which add terms to the classical advection-dispersion equations) must be fitted ("calibrated") by external software programs from data obtained from tracer experiments. This type of software is based on a "trial and error" procedure of the "Monte Carlo" type, which gives results with different settings, taking away the modeler's own ability to analyze.

The "Aggregate Dead Zone" ADZ method, unlike the previous method, TS, uses first-order differential equations that directly link the parameters of interest. Its transport mechanism is based on conceptually separating advection from dispersion, assigning each characteristic of these a corresponding volume. The total volume of flow (V) is assigned to advection, and the fully mixed volume (Vm) is assigned to dispersion, maintaining the model of net "separation" between different zones of the channel. This model ADZ also uses computational software resources to fit the results as the previous model.

4.4 Some dynamic and thermodynamic problems of these models: Maxwell's devil and null exergy for the process

In thermodynamics, "Work" is a macroscopic concept associated with the raising or lowering of a weight, the stretching or shortening of a spring [19], or any other means of transducing energy on a human scale, what O. Levenspiel [20] has called "axis work." Fortunately, when chemical imbalance is studied (which is the case of dispersion in a flow), it can be considered thermal and mechanical equilibrium since a fluid medium will have the same temperature and pressures, which greatly simplifies its analytical treatment. This treatment can focus almost exclusively on the nature of particle distributions and the conditions that determine mass transport.

As stated, while the longitudinal dispersion is essentially due to the shear rupture in the fluid, the transverse diffusion (in Y Axis) is almost entirely due to the turbulent mixing, which, in addition to having a Fickian nature, can be developed analytically from an initial discontinuity of the "Dirac impulse" type (**Figure 20**).

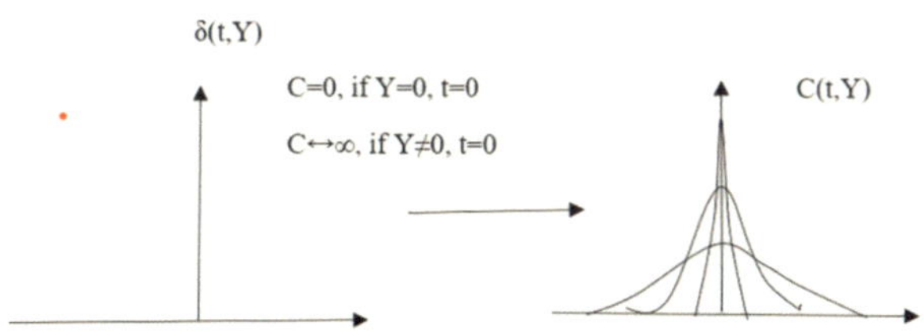

Figure 20.
Abrupt cuts in concentration tend to smooth behavior.

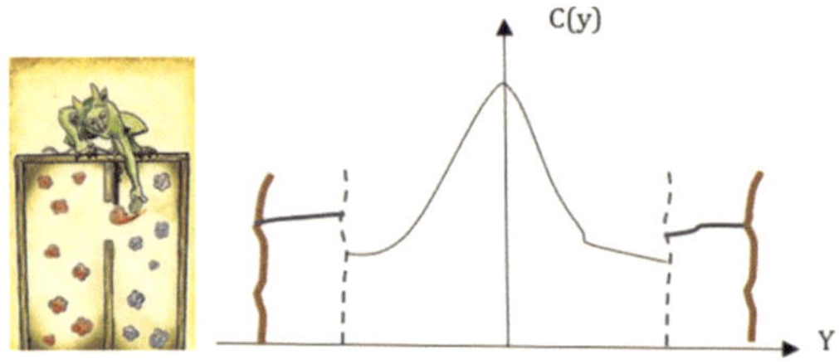

Figure 21.
Maxwell's devil choosing slow particles.

This development shows that, regardless of whether initial conditions of *discontinuity* are imposed, as would be the case with the Dead Zones (or Storage Zones) and the Central Zone), the dynamics should *always* become a *soft Fickian function*. Any arbitrary cut that is made on the distribution always leads to smooth, monotonous solutions.

A *non-natural* distribution with *abrupt changes,* such as those proposed by the TS or ADZ model in peripheral zones, would be shown as follows (**Figure 21**).

This unnatural distribution would only be possible if a "Maxwell's Demon" chooses only the *slow particles*, in the discontinuation of concentration, to store them. This, of course, would violate the 2nd Principle.

In this particular case, the *only* work that can be identified is that of frictional forces, so *all* the energy that goes into the system is spent on heat, and therefore the "available" energy for this *spontaneous process* of solvation-dissolution can be expressed as [20, 21]:

$$W(available) \approx T * \Delta iS \approx 0 \tag{19}$$

So, although the conceptual basis of this method is wrong, the operational time equations, obtained by software manipulation, are still useful in reactors and water quality studies.

5. A new vision of advection-diffusion. The state function that describes its evolution in turbulent flows

5.1 Nonlinear nature of molecular phenomena as a basic problem of physical models, and a macroscopic way of studying these processes

Nonlinearity, as an inevitable fact in physical reality, has been a permanent obstacle in the configuration of models of physical mechanisms [22], particularly those that attempt to describe molecular interactions in phase changes, such as those that occur in the formation of the tracer plume in a turbulent flow. In this process, there is a change from the solid state of the solute to a liquid state when it mixes with water.

The practical way to handle this problem has been to expand the nonlinear NS equation by means of numerical methods, whereas there are no general methods to solve it. This is because in this type of differential equation, main variables cannot be separated and cannot be identified separately [23]. This situation of "intractability" can be lessened if the "statistical sufficiency" condition of the system is met, and it is possible to identify and manipulate variables separately, such as thermodynamic flows and forces.

DOI: http://dx.doi.org/10.5772/intechopen.1004448

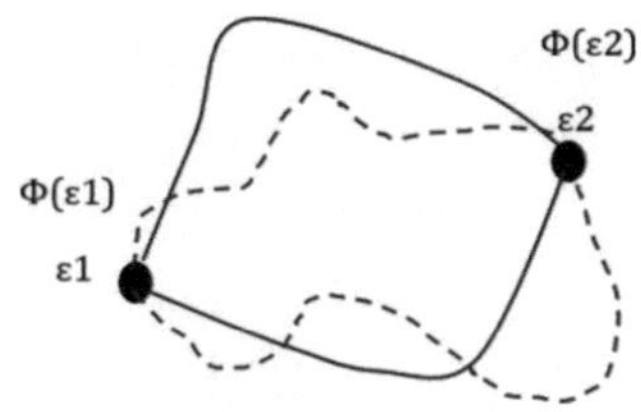

Figure 22.
Point function replacing nonlinear functions.

The microscopic non-predictability of the processes described by the nonlinear differential equation (that describes Poincare trajectories that collapse), can be overcome by a macroscopic thermodynamic analysis if the so-called "State Functions", or "point function", Φ(ε1,ε2) between two concrete states, ε1 and ε2, are used, since, as is known, despite the fact that the phenomenon is irreversible, nonlinear, (dotted lines) the evolution of this type of functions can be calculated by reversible (full lines) trajectories, in which the deviation from equilibrium is considered infinitesimal by devising ideal "quasi-static" processes as shown in **Figure 22** [3, 24].

$$\int_{\varepsilon 1}^{\varepsilon 2} d\phi = \phi(\varepsilon 2) - \phi(\epsilon 1) \tag{20}$$

This implies that, currently, although dynamic processes are "non-computable" at the differential level, "point functions" can be computed based on their initial and final macroscopic states. Nowadays [3], it is accepted that if temporal changes are used for the functions involved in the change from $\varepsilon 1$ to $\varepsilon 2$, under conditions of local equilibrium, it is feasible to calculate these functions on a "continuum" of time, restoring the analytical capacity on this type of events. State function then accomplishes the Schwartz condition if the system is on "statistical sufficiency."

$$\oint d\phi(t) \approx 0 \text{ if } W \ll RT \tag{21}$$

In this way, each pair of consecutive points of the "trajectories" is considered as the "beginning" and "end" of the elementary processes that compose it. In this way, a nonlinear evolution of the phenomenon can be represented by a continuous function that resolves such evolution.

5.2 Distributions in equilibrium. Svedberg's constant in experiments on Brownian motion

In the first "linear" regime described in 2.2 [25, 26], thermodynamic equilibrium is basically characterized by the "Principle of detailed equilibrium" in which it is stated that "each identifiable physical process must be counteracted separately by its opposite" and is characterized by the independence (no interaction) of events, as established by the Poisson distribution. Such a representation is assigned to the initial observations of Brownian motion, as is analyzed below.

At the beginning of the twentieth century, the Swedish physicist T. Svedberg, in his investigations on the movement and distribution of colloidal particles seen under a microscope in an ultracentrifuge machine, verified the "Brownian motion" of such particles (**Figure 23**).

Figure 23.
Svedberg in front of his equipment [26].

Number of particles, m	Cases of occurrence	Relative frequency	Poisson's probability
0	112	0.216	0.214
1	168	0.325	0.330
2	130	0.251	0.254
3	69	0.133	0.131
4	32	0.062	0.050
5	5	0.010	0.016
6	1	0.002	0.004
7	1	0.002	0.001

Table 1.
Experimental and theoretical results of Svedberg observations.

These experiments, made on the count of colloidal particles on lattices observed from time to time, yielded very interesting numerical data, which were summarized statistically, in **Table 1**.

The observed average value of this distribution, of # particles per unit of grid and per unit of time, <a>, is calculated as follows:

$$<a>\approx\frac{0*112+1*168+2*130+3*69+4*32+5*5+6*1+7*1}{518}\approx\frac{801}{518}\approx 1.546 \tag{22}$$

This is the so-called "Svedberg number" [27], based on the distribution observed in the first two columns, and with the probabilities of occurrence, both from the combinatorial calculus (binomial distribution) and from the Poisson distribution, in accordance with the equilibrium (independence) assumption of the phenomenon.

5.3 Natural growth, geometric progression, and Brownian motion

For the Brownian motion, the total transport time will be described by the sum of a convergent series determined by the following differential equation, named "logistic curve, a nonlinear pattern with a development region and a saturation. This dynamic corresponds to a geometric convergence scaling, as in **Figure 24** [22].

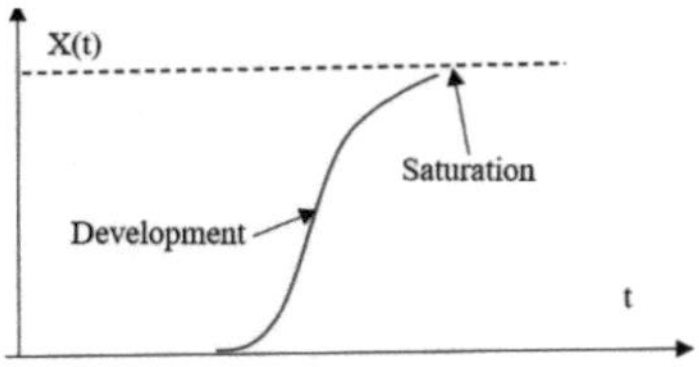

Figure 24.
Physical "law of growth."

Focusing only on the "development" part, in which the growth differential is proportional to the growth itself, you can write:

$$\frac{dr}{r} \approx \int k * f * dt \tag{23}$$

Where "r" is the ratio between sequential quantities related to this transport, the solution of which is:

$$\mathrm{Ln}(r) \approx \int k * f * dt \tag{24}$$

The definition of the type of function "*f*" that is involved depends on the phenomenon that is being studied, for the case of Brownian motion, that function "f" would be a concentration of a solute, and "*k*" is the inverse of a remarkable time, T, we have for this the mean value $< C >$ of Svedberg's experiment, but in another context.

$$r(t) \approx e^{\frac{1}{T}\int C(t) * dt} \approx e^{<C>} \approx e^{1.54} \approx \left(1 + <C> + \frac{<C>^2}{2!} + \dots + \frac{<C>^n}{n!}\right) \tag{25}$$

For the integral of (22), the ratio "r" corresponds to the division t/τ, where τ, as we know, is the basic (microscopic) time of the transport process and t is the total observation time. This solution is also identified with the "scale invariance," hallmark of the Neperian function, implying a geometric growth over the value 1.54.

$$t \approx \tau\left(1 + 1.54 + \frac{1.54^2}{2!} + .. + \frac{1.54^n}{n!}\right) \approx \tau * 4.665 \tag{26}$$

The value $\delta \approx 4.669$ appears as a "Universal Constant" in the work of M. Feigenbaum in 1975, [28] applying the concepts of "renormalization" that had been successful in the interpretation of nonlinear phase changes, understood that this technique, based strongly on the self-similarity of these phenomena, insofar as their description at different levels is mathematically identical, could be useful to apply to turbulence, in which there is a cascade of vortices that are included equally, one within the other. Considering an infinite process of bifurcations (period bending) for monomodal descriptions, strongly associated with the "logistic curve," which describes in a simple way the processes of growth with losses.

Feigenbaum, for this case, found a "bifurcation tree" which, a constant of geometrical progression for the remarkable times, k, was coincident with that of Svedberg ratio:

$$\delta = Lim\frac{k_n - k_{n-1}}{k_{n+1} - k_n} \approx \frac{\Delta k(previous)}{\Delta k(posterior)} \approx 4.6692 \tag{27}$$

5.4 Molecular basis of turbulence. Einstein's mechanism and Svedberg's constant

In his classic work on Brownian motion, A. Einstein [29] defined the sequential motion of molecular plots that move elementary distances, Δ, in a characteristic time, τ, in such a way that the mass transport described is in the condition of "statistical significance," maintaining the mean thermal energy of dynamics.

Then, a fixed Eulerian observer (blue) on the bank of a river, who measures the diffusion of a solute, will describe that as time passes, the arrival of sequential groups of solute molecules, each with a given concentration, n, as a function of time θn (**Figure 25**).

When the observer has done all the counting of the tracer particles, and the relationship (24) is established, it is possible to define a "diffusion velocity" such that: [30].

$$vd \approx \frac{\Delta}{\tau} \tag{28}$$

This expression can be extended according to Einstein's definition, with D as the diffusion coefficient:

$$vd \approx \frac{\sqrt{2D * \tau}}{\tau} \approx \sqrt{\frac{2D}{\tau}} \tag{29}$$

Note that in this definition we should use not time as the general variable, "t," but the "characteristic time," τ, since we are describing molecular proper motions in the condition of "statistical significance," which must, by definition, be shorter than the independent variable.

5.5 The state function Φ (U,D,t) that describes the evolution of the solute in the turbulent flow: the modified advection-diffusion equation

Since the phenomenon of dispersion depends to a large extent on the breakdown effect generated by advection, inversely proportional to the mean velocity, U, an evolution function, Φ *(U,D,t)* can be defined as follows:

$$\Phi \approx \frac{vd}{U} \approx \frac{\left(\frac{\Delta}{\tau}\right)}{U} \approx \frac{\sqrt{\frac{2D}{\tau}}}{U} \tag{30}$$

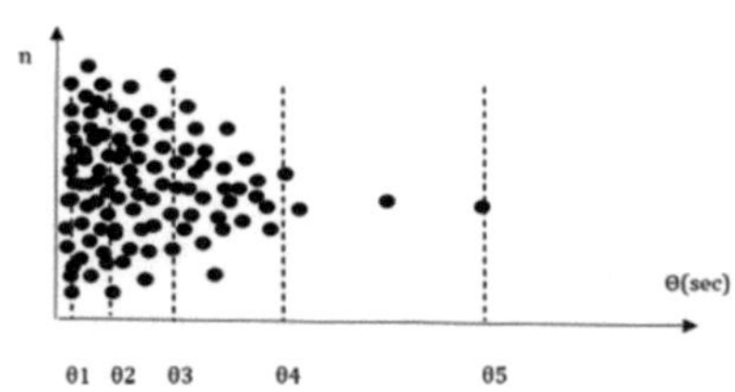

Figure 25.
Observer counting time arrivals of groups of tracer particles.

This function can also be developed as shown, with σ_t as the temporal variance of the *Tl* distribution of the tracer. Here $2.16 \approx \sqrt{\delta}$.

$$\Phi \approx 2.16 * \left(\frac{\sigma_t}{tp}\right) \tag{31}$$

From Eq. (30), you can find a modified version of Fick's basic definition, nonlinearity (16), which fulfills representing the experimental tracer curves, without using the fallacious arguments of the *TS* or *ADZ* method:

$$C(t, X) \approx \frac{M}{Q * \Phi * t * 1.16} e^{-\frac{(tp-t)^2}{2*0.214*(\Phi*t)^2}} \tag{32}$$

From this equation, with the solute mass, taking the peak value of the concentration and time, and the state function for that time, we can calculate the flow rate, *Q*:

$$Q \approx \frac{M}{Cp * \Phi * tp * 1.16} \tag{33}$$

Also, using the modified Fick function, but as a function not of time but of distance, the cross-sectional area of the flow, *Ayz*, can be calculated with the plotter, using a multiplier by 1000 in the denominator to fit the units.

$$Ayz \approx \frac{M}{1000 * \int_{x1}^{x2} C(X) dX} \tag{34}$$

That is, it can be applied to describe nonlinear phenomena, such as turbulence, without having to solve the analytical problems associated with the NS equation. The curve representing Φ as in **Figure 26** is proportional to the curve of the thermodynamic evolution of solute advance in **Figure 16**.

In the dispersion stage, both solute ions and solvent (water) molecules are mixed in flow, and distributed in ever-increasing volumes, until when $\Phi \approx 0.38$, condition of "complete mixing" (homogeneous distribution of the tracer in the cross-section of the flow), when the velocity of the ions is almost that of the flow itself, according to **Figure 17**.

This is important because it allows you to have an analytical function that describes turbulence, without resorting to NS-type functions.

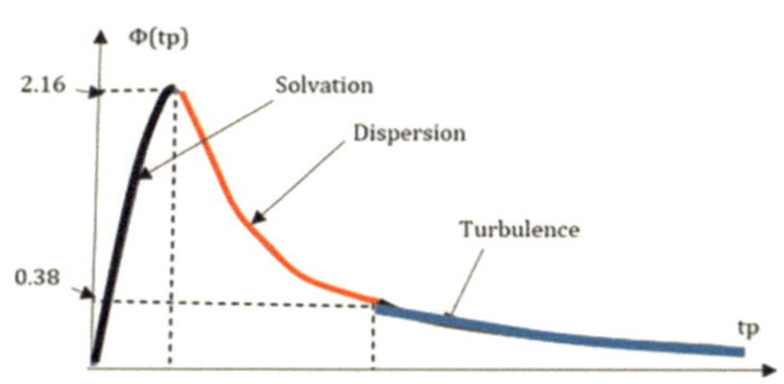

Figure 26.
State function curve as behavior curve.

A practical definition of the state function, which is useful for calculations in the turbulence stage is as follows, with "*M*" the mass of the tracer, "*Q*" the discharge, "tp" is the peak time of solute distribution, and "*γ*" a constant that depends on the type of solute:

$$\Phi \approx \frac{M}{Q * \gamma * 1.16} * \frac{1}{\sqrt[3]{tp}} \tag{35}$$

Also, it is valid the following equation:

$$Cp \approx \gamma * tp^{-\frac{2}{3}} \tag{36}$$

Using these equations, we can write the following equation:

$$\Phi \approx \frac{M}{Q * \gamma^{3/2} * 1.16} * \sqrt[2]{Cp} \tag{37}$$

This result is coincident with the well-known behavior of ejected exothermal heat of solute in **Figure 16**.

On the other hand, one can, of course, equate the one-dimensional CM definitions (8) with Newtonian interactions, and the relation from Φ with Van-der Waals (electrical) dispersal interactions.

$$\frac{R^{\frac{2}{3}}}{n} * \sqrt{S} \approx \frac{1}{\Phi} * \sqrt{\frac{2D}{\tau}} \tag{38}$$

From this relationship, it is possible to establish the hydraulic variance as a function of the dispersive variance, and also calculate both hydraulic and dispersive magnitudes knowing values of Φ. This paves the way for the measurement of large streams, which is impossible to do with the state of the art.

$$\sigma(hydraulic) \approx f(\sigma(dispersive)) \tag{39}$$

In this way, the state function allows you to look precisely into the dynamics of turbulent flow, insofar as it can be calculated, *U*, *Q*, *Ayz*, y *D*, which is facilitated by the "Dynamic Equilibrium" condition of the channel, which presents large volumes in the condition of "Coarse grained" with nonlinear "patches," which can be represented by σ_u (**Figure 27**).

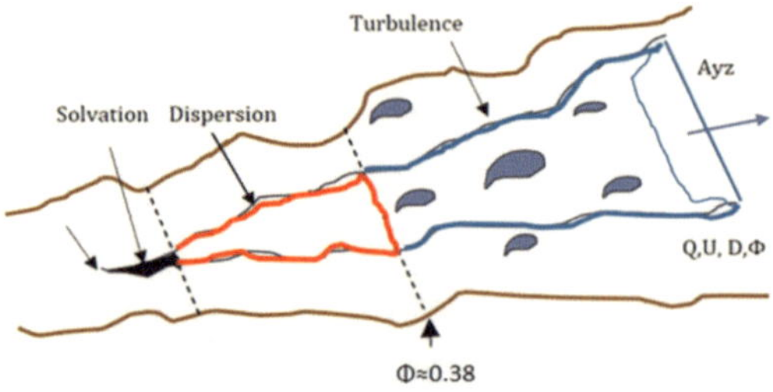

Figure 27.
State function allows large river measurements.

6. Power function from the state function, which describes the ratio of notable numbers in dispersion and turbulence

It is interesting to see how the notable Feigenbaum numbers and the Golden Number are distributed in a certain order with respect to dispersion and turbulence, as power function of $\Phi\ (k)$ (**Figure 28**).

This arrangement can be interpreted in that once dispersion begins, a relationship is established between the characteristic time, τ, and the general time, tp (by means of δ). Then, the relative, sequential vigor of the fluctuations is determined by the constant α. Finally, the rotational dynamics of the turbulence are established by the Golden Number, ψ, which establishes the vortices by Fibonacci scaling [31, 32].

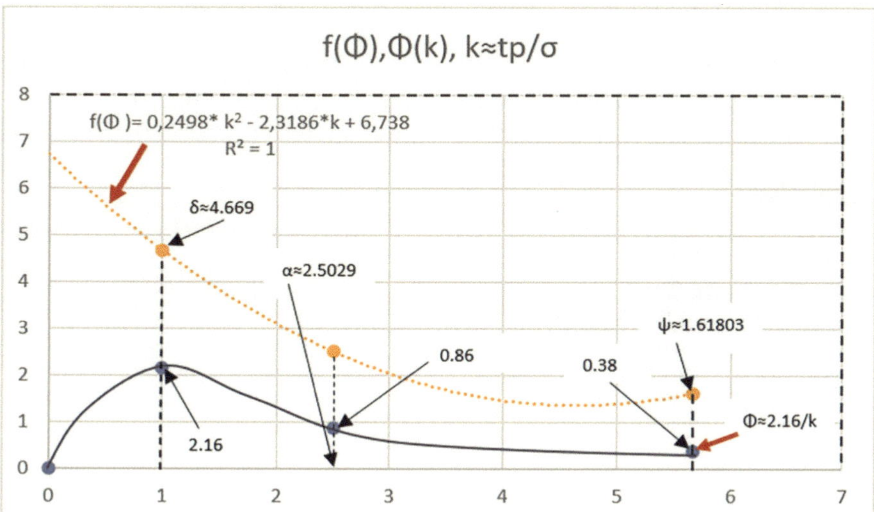

Figure 28.
Power function of $\Phi(k)$ [31].

7. Conclusions

1. The physical model analyzed here shows that turbulence, for practical purposes, can be looked at from a different point of view than the reductionist scheme of nonlinear partial differential eqs. (NS).

2. Tracers are fluorescent marker salts, which, when dissolved in natural flows, allow the characterization of heterogeneous movements of water, since these movements are the same as those of the solute molecules.

3. The great fluidity of water means that the usually "violent" movements of turbulence can be encased in the "linear" region of irreversible thermodynamics, where the "forces" (gradients) and the "flows" (rates of mass transport) are proportional, and the condition of "statistical sufficiency" is maintained, thanks to the fact that the "activation energy," ΔW, is much less than thermal energy, RT.

4. It is shown how the use of "thermodynamic potentials," in this case, a function of state, Φ, which is proportional to the enthalpies put into play, of "solvation" (when injecting the tracer), and of "dilution" as this cloud evolves, it is possible to obviate the difficult application of the NS equation, to obtain useful data in natural channels that are in "Dynamic equilibrium".

5. In the definition of Φ, the "Svedberg constant," $a \approx 1.54$, which is the fractal hallmark of diffusive, Brownian, processes, plays an important role, allowing the calculation of the main variables in all types of channels.

6. In this paper, it is shown that the state function is proportional to square root of concentration, $\sqrt{C}$, as is expected from classical thermodynamics of solute evolution in flow.

7. It is possible to establish a power function that encompasses a sequence of the remarkable numbers that govern the fractal nature of turbulence.

Acronyms list

NS	navier-stokes
Re	reynolds number
Rec	critical reynolds number
Pr	prandtl
CM	chezy-manning

Author details

Alfredo Constaín Aragón
Fluvia Tech, Bogotá, Colombia

*Address all correspondence to: alfredo.constain@gmail.com

References

[1] Annila A, Salthe S. Physical foundation of evolutionary theory. Journal of Non-Equilibrium Thermodynamics. 2010;**35**:301-321

[2] Annila A, Pernu T. Natural Emergence. Vol. 17, No. 5. Berlin: Wiley Periodicals. Inc; 2012

[3] Kondepudi D, Prigogine I. Modern Thermodynamics: From Heat Engines to Dissipative Structures. Chichester: Wiley; 1998

[4] Annila A, Sharma V. Natural process-natural selection. Biophysical Chemistry. 2007;**127**:123-128

[5] Frenkel J. Kinetic Theory of Liquids. New York: Dover; 1955

[6] Nekrasov B. Hidraulica. Moscú: Editorial Mir; 1955

[7] Stewart I. 17 ecuaciones que cambiaron el mundo. Barcelona: Critica; 2013

[8] Vennard J. Elementos de la mecánica de fluidos. México: C.E.C.S.A; 1965

[9] Annila A, Makela T. Natural patterns of energy dispersal. Physics of Life Reviews. 2010;**7**:477-498

[10] Anner G. Elementary Nonlinear Electronic Circuits. Englewood Cliff: Prentice Hall; 1967

[11] Penrose R. Ciclos del tiempo. Barcelona: Debate; 2010

[12] Ochoa T. Hidraulica de ríos y procesos morfológicos. Bogotá: ECOE; 2011

[13] Leopold L. Downstream change of velocity in rivers. American Journal of Science. 1953;**251**:606-624

[14] Damaskin B, Petri O. Fundamentos de la electroquímica teórica. Moscú: Mir; 1981

[15] French R. Open Channel Hydraulics. New York: McGraw-Hill; 1985

[16] Fischer HB. Longitudinal dispersion and turbulent mixing in open channel flow. Reviews of Fluid Mechanics. 1973; **5**:59-78

[17] Hart D. Parameter estimation and stochastic interpretation of the transient storage model for solute transport in streams. Water Resource Division. 2000; **31**:1-21

[18] Lees M, Camacho L, Chapra S. Modeling solute transport in rivers under unsteady flow conditions: An integrated velocity conceptualization. In: BHS 7th National Symposium, 21–29, 2000. New Castle, UK: New Castle University; 2000

[19] Semansky M. Calor y termodinámica. Madrid: Aguilar; 1968

[20] Levenspiel O. Fundamentos de Termodinámica. New Jersey: Prentice Hall; 1998

[21] Planck M. Treatise on Thermodynamics. New York: Dover; 1945

[22] Stewart I. Juega Dios a los dados? Barcelona: Grijalba-Mondadori; 1991

[23] Annila A, Ketto J. The capricius character of nature. Life. 2012;**2**(1): 165-169. ISSN 2075-1729

[24] Sonntang R, Van Wilen G. Introduccion a la termodinámica clásica y estadistica. México: Limusa; 1995

[25] Von Mises R. Probability, Statistics and Truth. New York: Dover; 1957

[26] Una Vuelta por la Suecia de 1920. Available from: https://www.beckman.es/resources/technologies/analyticalultracentrifugation/history/spin-through-sweden

[27] Constain A, Mesa D, Peña C, Acevedo P. Svedberg's number in diffusion processes. In: Advances in Environmental Sciences, Development and Chemistry. The 2014 International Conference on Water Resources, Hydraulics & Hydrology Proceedings, Santorini, Greece. 2014

[28] Gleick J. Chaos: The Making of a New Science. New York: Mc Graw-Hill; 2008

[29] Einstein A. Investigations on the Theory of the Brownian Movement. New York: Dover; 1956

[30] Constain A. Definición y análisis de una función de evolución de solutos dispersivos en flujos naturales. Dyna. 2012;**175**:173-181

[31] Constain A, Peña G, Peña C. Función de estado de evolución de trazadores, Φ (U,E,t) aplicada a una función de potencia que describe las etapas de la turbulencia. Vol. 199. Madrid: Revista Ingenieria Civil; 2021

[32] Livio M. La Proporción aurea. Barcelona: Ariel; 2002